ARCHANGEL

THE
AMERICAN WAR
WITH RUSSIA

BULKINGTON BOOKS

"Madman! Look through my eyes if thou hast none of thine own."

This Book Belongs to:

SA
PIEN
TIA

SCI
EN
TIA

IN
TELLEC
TVS

FOR
TITV
DO

CON
SILI
VM

PI
E
TAS

TIMOR
DÑI

Americans on patrol in the snow in north Russia.

ARCHANGEL
THE
AMERICAN WAR
WITH RUSSIA

By A Chronicler, Ambassador John Cudahy

Nothing extenuate, nor set down aught in malice.
— OTHELLO

A. N. R. E. F.
1918 - 1919

Polar Bear Monument, dedicated to the A.N.R.E.F., in White Chapel Cemetery, Troy, Michigan. Sculpted by Leon Hermant.

Sergeant William H. Bowman,
339th United States Infantry

TABLE OF CONTENTS

Publisher Note

Blessed beloved bookreader, you have found this volume in your vision. We hope you read on, but let us offer a few humble words. Of making many books there is no end, and a long preface is a chasing after wind. We pray you give us a moment's indulgence.

Bulkington Book's mission is to build a bridge into the past, before film, television, copyright, and internet swallowed up the world. Before 'content' was king of the culture. Before slop was the aim. Before brainrot was the goal. If the reader enters the portal into the past and finds friends from before the echo chamber, they may find armor and sword against the dreadful noise machine. The reader will remember and honor their heritage and humanity.

We are convinced that many authors and many books are ready to rise like Lazarus and re-enter the world to remind the readers that their life has purpose; that their time should be valued; and that it is happiness to honor your heritage.

We found this story worthy of revival, and we hope you find it worth your while.

Your Most Humble & Obedient Servant (YMHOS),

Arthur Bulkington,

Melville Bay

[One note of apology. The original text included several photographs from the front. However, owing to the poor archival scans, we have omitted some of them in this republication. To make up for those omissions, we have gathered many other photographs from the time period and interspersed them in the book.]

PUBLISHER'S FOREWORD

Ambassador John Cuhady's account of his time in North Russia is straightforward. We have supplied footnotes for some context about the names and terms he uses. You can read on without the aid of this foreword if you so desire.

But since we have taken it upon ourselves to reissue this short firsthand account of the American North Russian Expeditionary Force (A.N.R.E.F), we feel that we owe the Reader a few words on why this was worth our time—and why it is worth yours.

We have three points to make:

1) We will tell you why we liked it, why it was worth our time to read, reformat, and republish.
2) We will suggest reasons why this book is relevant for our time.
3) We will suggest why it may be worth your precious reading time.

Let's start with 1). We liked it because it is an honest firsthand account. Firsthand accounts reveal more than official histories. Real truth in history gets ironed out to make facts and events serve later narratives.

This fact is one of the purposes of our work here at Bulkington Books. We believe that modern culture is "stuck." (This point has been widely popularized by a Mr. Paul Skallas, a.k.a. 'Lindyman,' "stuck culture," but many others have expanded on his initial insight.) Culture stuck inside of a copyright echo chamber of intellectual property—reboots, sequels, spinoffs, shallowness and slop. Almost all this 'content' is set in the pocket universe of Postwar American Triumphalism. Culture is stuck inside a Saturday morning cartoon morality tale that justifies foolish wars in Iraq and Ukraine. Or worse, culture is stuck inside a 30 second clip of dancing girl soldiers priming the viewer to support war. We submit that breaking out of the pocket universe of Postwar American Triumphalism, and back into the open steppes of deep history, is an eminent good. It will dissolve shallow discourses, shatter metanarratives, and sidestep the received court history, freeing curious minds to roam back into the centuries.

The disastrous episode of the undeclared war in Russia is a perfect case study—servicemembers died in vain in an undeclared war. Their memory goes unhonored, and the lessons of this venture go unlearned. And more have—and will—suffer the same fate.

All that is to say, when we find a first person account written by a free man telling the truth about a forgotten corner of history, we feel we are breathing in clean air and drinking pure water.

The author, John Cudahy, has an axe to grind here. This is why he wrote anonymously. He does not hide his feelings that the whole venture was doomed from the start. It lacked sufficient resources and strategy and vision.

Only a fool would think that a few thousand men in North Russia could link up with a few thousand men from Vladivostok. To illustrate, it would be as if a few thousand Russians landing in Spokane, and a few thousand Russians landing in Savannah, would be able to link up and conquer the United States. That is what the policymakers in the Wilson Administration thought would happen when they went into Russia.

We are living in an age where history is written second, third, fourth hand. Academics don't read primary sources, they respond to prior academics about the 'meta' narrative. This tendency bedevils most modern scholarship. The event doesn't matter, the details don't matter, the interpretation of the event, in furtherance of modern ideology, in response to other commentators, is what matters. Then people take the fourth-hand crumbs of history they got from this game of telephone and use this to form their opinions.

But if you were sent to fight for your country, you weren't told why, and there was no obvious reason, no sane objective, and there wasn't enough resources or men to do the job, you're not going to tolerate the fourth hand accounts you hear from people who want to warp the facts to their ends.

John Cudahy is angry that his country sent him and his companions over to North Russia to fight and die defending a frozen hellhole. He was so disillusioned with the venture that he wrote this book anonymously, in a fit of rage. This is a good reason to republish a book. To honor this man's rage. America could very well make the same mistake again. We hope Mr. Cudahy finds readers who can prevent this from happening.

In short, we liked it because it is raw. It exposes a forgotten episode of history. It tells truth.

We like an angry honest man warning against disastrous wars. We also liked it because he is a good writer. He rises to the occasion and touches upon the eternal things. He quotes poetry. He bends his own prose upwards to the heights. He needed to find the words because he needed to honor the men he fought with. He could not move on in silence. He would not forget.

Coming to point 2), there can be no subtlety here. The United States invaded Russia in 1918 without declaring war and wasted blood and treasure to no good effect. The United States has puppeted Ukraine and used it to wage war on Russia, since at least 2022 if not 2014, without declaring war, at great cost of blood and treasure, with no good effect.

Over a hundred years ago, as Russia was falling apart after taking millions of casualties, the United States decided to invade their former ally by landing troops in Archangel and Murmansk, with British, Canadian, and other detachments. American doughboys collaborated with Japanese troops in the Russian Far East. British forces also worked up from the South through Persia and the Caspian Sea. America and the other Allies assisted White Russian forces of all stripes, monarchists, non-Bolshevik socialists, liberals, and democrats. The unintended consequence was that these incursions and alliances forced the Russian people to rally behind the Bolsheviks, and united them against all these foreign enemies.

In our time, the ceaseless pressure on Russia from all angles has actually united the Russian nation behind Putin.

The current conflict has reinvented World War One style

trench warfare. In World War One, new military technology, advances in artillery and machine guns, imposed a stalemate status quo on the front. Cavalry was rendered obsolete. The use of drones in the current conflict has recreated this trench line stalemate and severely limited the utility of armored formations in maneuver warfare. Breakthroughs and territorial gains come at a very high cost.

The Russian Civil War has echoes in the current war. Many of the heroes of modern Ukraine grew out of the aftermath of the First World War. White Russian forces would have taken Moscow had not Trotsky made a tactical alliance with the anarchist forces under Nestor Makhno. Strange political bedfellows is nothing new in Ukraine.

The Biden Administration in Ukraine, and the Wilson Administration in Russia deserve comparison. Neither of them spelled out a clear endgame. What would Woodrow Wilson do if the American troops in the North had been more successful? Would he have restored the Monarchy, and helped set up a new Tsar? Would he have imposed a new government upon Russia as new governments were imposed on Austria, Yugoslavia, Czechoslovakia, Poland, et al?

Likewise, one must wonder what the aims of the Biden administration are in Ukraine. Would it be over if Zelensky marched into Moscow? What would be the benefit of a Russian President hand picked by the State Department? If they achieved a total victory, what would that actually result in? What would it look like?

It seems like the dog wouldn't know what to do if it caught the car.

The boogeyman of Bolshevism is no more. There is no threat of world revolution or class warfare. The unions are weak. There is no Internationale. President Vladimir Putin, for all his bad deeds and demonization, cannot fill the role of Bolshevik boogeyman like Lenin and Trotsky did.

From a purely military point of view, the Wilson administration relied on half measures. It did not send enough men or equipment to North Russia or to the Far East. It did not commit entirely to siding with the White Russians. It was caught between sound military strategy and political calculus. And the cost was compromise and half measures and ultimately pointless bloodshed – and for those left behind in North Russia, or the Russian Cossacks, betrayal and abandonment. The Biden administration seems to have made the same mistake. It has not committed fully to total victory, but instead has relied on half-measures and piecemeal commitments. The price tag is much higher, but the ultimate reluctance to go 'all-in' is the same choice for the same reason.

A man's reach exceeds his grasp. Wilson could reach Archangel and Vladivostok, but he could not grasp them. Likewise, Biden can reach into Ukraine and deep into Russia with the forces he is funding and supporting, but neither he nor Zelensky seem able to fully grasp the conflict.

The American conflict with Russia is very old. The Russians have a long memory. The infamous monologue Putin gave in his interview with Tucker Carlson, marshaling a thousand years of Russian history and identity, confounded western audiences trapped in the pocket universe of modern culture. With a grasp of the parallels of the A.N.R.E.F. expedition and the Special Military Operation in Ukraine, the Reader can reach past the bloodthirsty armchair generals online and see the real balance of power.

This brings us to point 3), why this might be worth your time as a Reader. To grasp the truth about American conflict with Russia, this book will serve you well.

Discussion of the current conflict is riddled in hubris and ideology. You can find hours of footage where you watch men fight to the death in hand to hand combat, or perish from a drone's explosion. People relish the hellish deaths of strangers if it helps them score points in a pointless debate online. Foreign policy has become wed to American culture war flashpoints. Transgender advocates defend Ukrainian Nazis like Azov Faction. The old anti-war policy positions of Democrats circa 2006 has been entirely replaced by a jubilant jingoism. The old war drums of the Republicans circa 2006 has been replaced with dovish de-escalation. Ukraine's borders are sacrosanct and the southern border of the United States is an open door. The boundaries of race and gender are infinitely fluid and permeable, but the Ukrainian political identity is written upon the Rock of Ages.

If the choices of the Wilson Administration were self-evidently foolish and shortsighted in their own day, the scope and scale of the foolishness of current policymakers seems positively reckless.

If a time traveler went back to Versailles, and showed them

documentation of the recent conflict in Ukraine, and told them they needed to make different choices, what would they do differently?

The incoming president, Mr. Donald Trump, has promised a peace deal. What kind of peace deal would prevent a future conflict? A generation of peace without fundamental resolution only gives the next generation a list of grievances as premise for the next war.

If a time traveller from a century hence were to come back, what would they show us? If one of us was able to go through a portal into the next century and see the results, what would the judgment of history be?

If these questions carry depth and weight for the Reader, then reading this narrative is worth the Reader's precious time.

For John Cudahy, writing after the end of the war, and the foolish excursion into North Russia, and the embarrassing retreat, there was no peace. He left men behind, and he knew their sacrifice was for naught. And he wrote this book to give voice to his outrage, and to warn the future.

If you believe that being informed about forgotten history is a sword and shield against foolish wars in frozen hellholes, then his warning, this book, is worth your time.

Graves of first three American soldiers killed in action on the North Russian Front. September 16, 1918. Left to right: Ignacy H. Kwasniewski, Mechanic, Co. 1; Anthony Soczkoski, Pvt, Co. 1; Phillip Sokol, Pvt. Co. 1; all of 339th Regiment Infantry.

American Key Men in European Peace

By a Staff Artist of The Christian Science Monitor.

Author Biographical Note

John Clarence Cudahy was born December 10, 1887, in Milwaukee Wisconsin. He was the son of Patrick Cudahy, an Irishman who emigrated to America and worked his way up the meat packing industry to become the owner of a factory.

John Cudahy attended Harvard in 1910, then earned his law degree at the University of Wisconsin. He joined up and was commissioned a lieutenant in the US Army. He was assigned to the 339th Regiment, composed of draftees from Michigan and Wisconsin, commonly referred to as "Detroit's Own." The 339th took on the nickname 'Polar Bears' when they were assigned to join a detachment of British and Canadian forces in North Russia.

Cudahy's service in the 339th in Archangel, Russia was exemplary. Though he minimizes his own conduct in the book to follow, he saved many men by leading a detachment fighting out of an encirclement by the Red Army.

He wrote this book, originally under a pseudonym, because he was dissatisfied with the way the Russian Expedition was organized and under resourced.

After the war, he returned home and took over the family business. He wrote several other books, about a trip to Mexico, and a safari to Africa. He was involved in the Wisconsin Democratic Party, and an early supporter of FDR.

In 1933, he was appointed the American Ambassador to Poland. From 1937 he was Ambassador to the Irish Free State. In 1940, he had a short stint as the Ambassador to Belgium and Luxembourg, which ended in May 1940 when they were invaded and had to form governments-in-exile.

He died in 1943, when he was thrown from his horse while riding on his estate in Wisconsin. He was survived by several children, including the entrepreneur and founder of Marquette Electronics, Michael Cudahy (1924-2022).

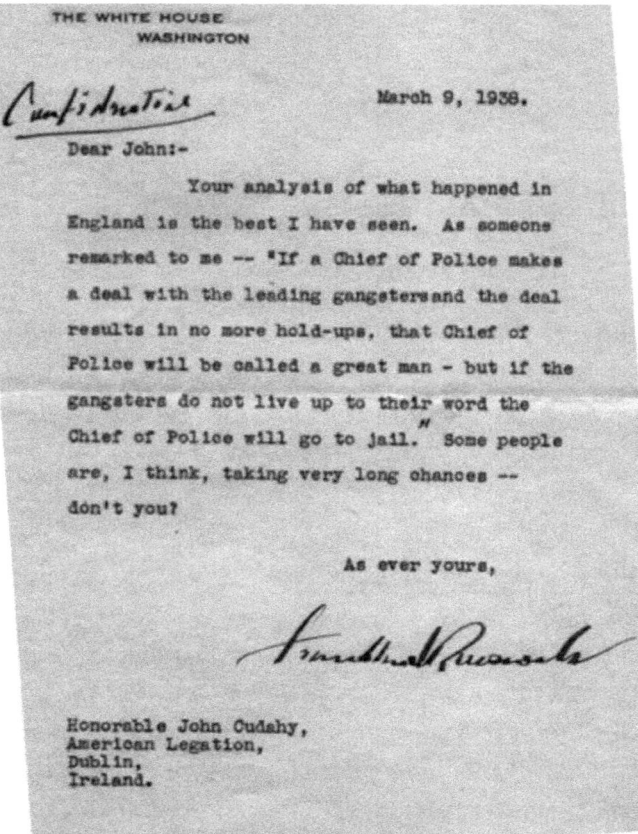

THE WHITE HOUSE
WASHINGTON

Confidential   March 9, 1938.

Dear John:-

        Your analysis of what happened in England is the best I have seen. As someone remarked to me -- "If a Chief of Police makes a deal with the leading gangsters and the deal results in no more hold-ups, that Chief of Police will be called a great man - but if the gangsters do not live up to their word the Chief of Police will go to jail." Some people are, I think, taking very long chances -- don't you?

                        As ever yours,

                        Franklin D Roosevelt

Honorable John Cudahy,
American Legation,
Dublin,
Ireland.

Did we declare war upon Russia
when we took a hand in the game
I know that we hopped onto Prussia
And Austria got the same.
But still I have no recollection
Of breaking with Russia, I swear
And cannot help making objection,
To having our boys over there
What quarrel have we with that nation?
Just how did it tread on our toes?
— *"What About Bringing Them Home,"* poem by
Michigander George Smith, 1919.

Professor E.K. Voss, from the author,
John Cudahy, Milwaukee, August 31 1930 (?),
from the manuscript used for this republication.

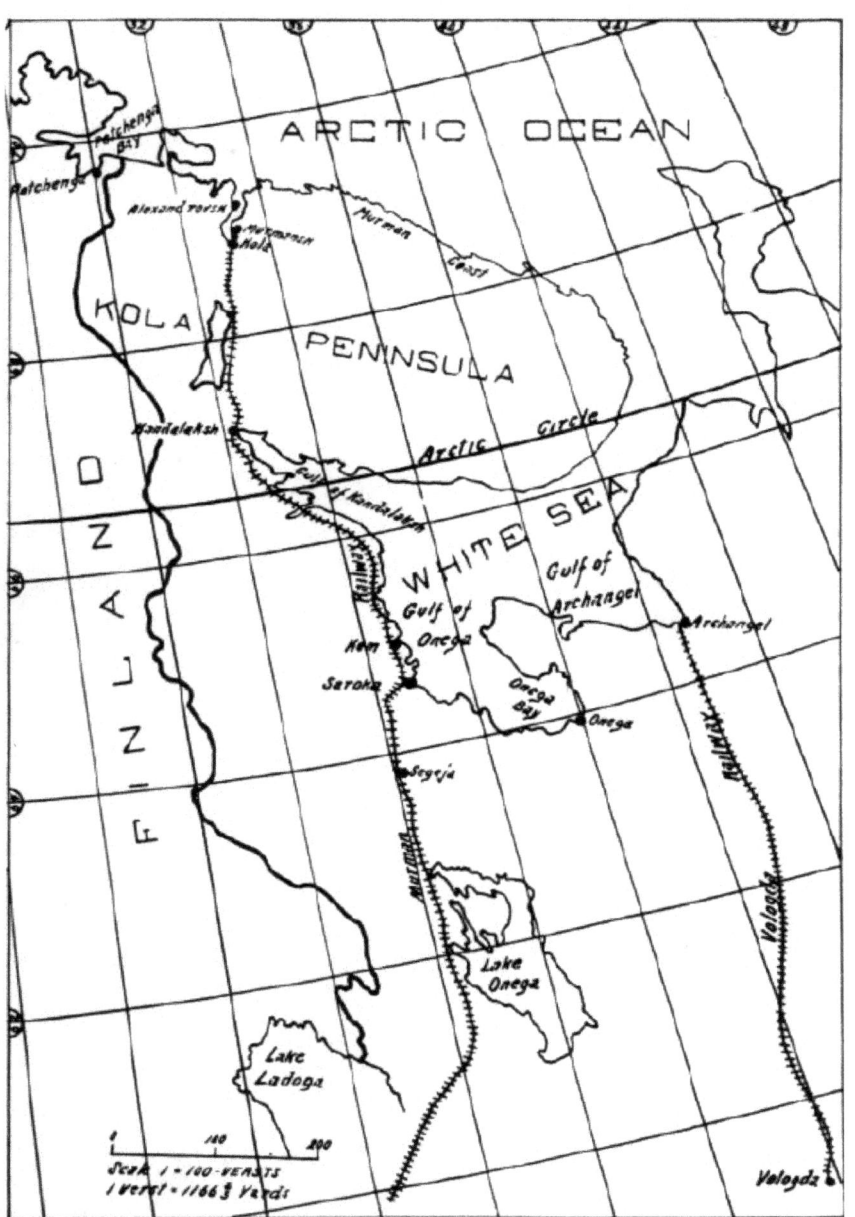

The Murman and Vologda railways

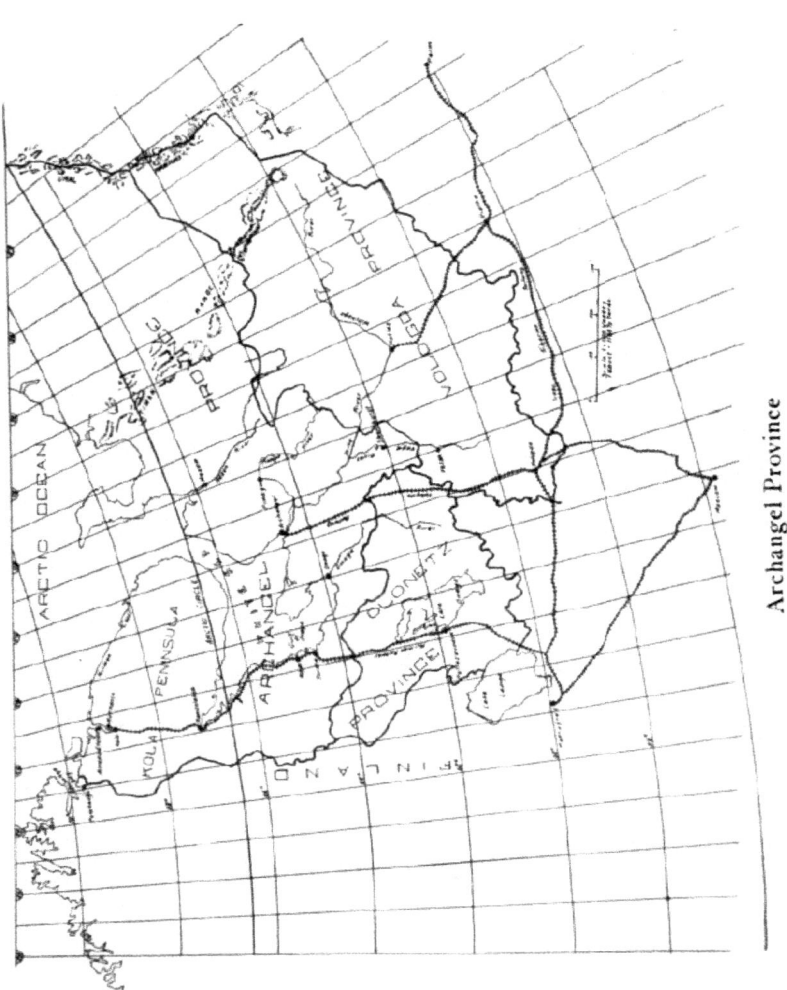

Archangel Province

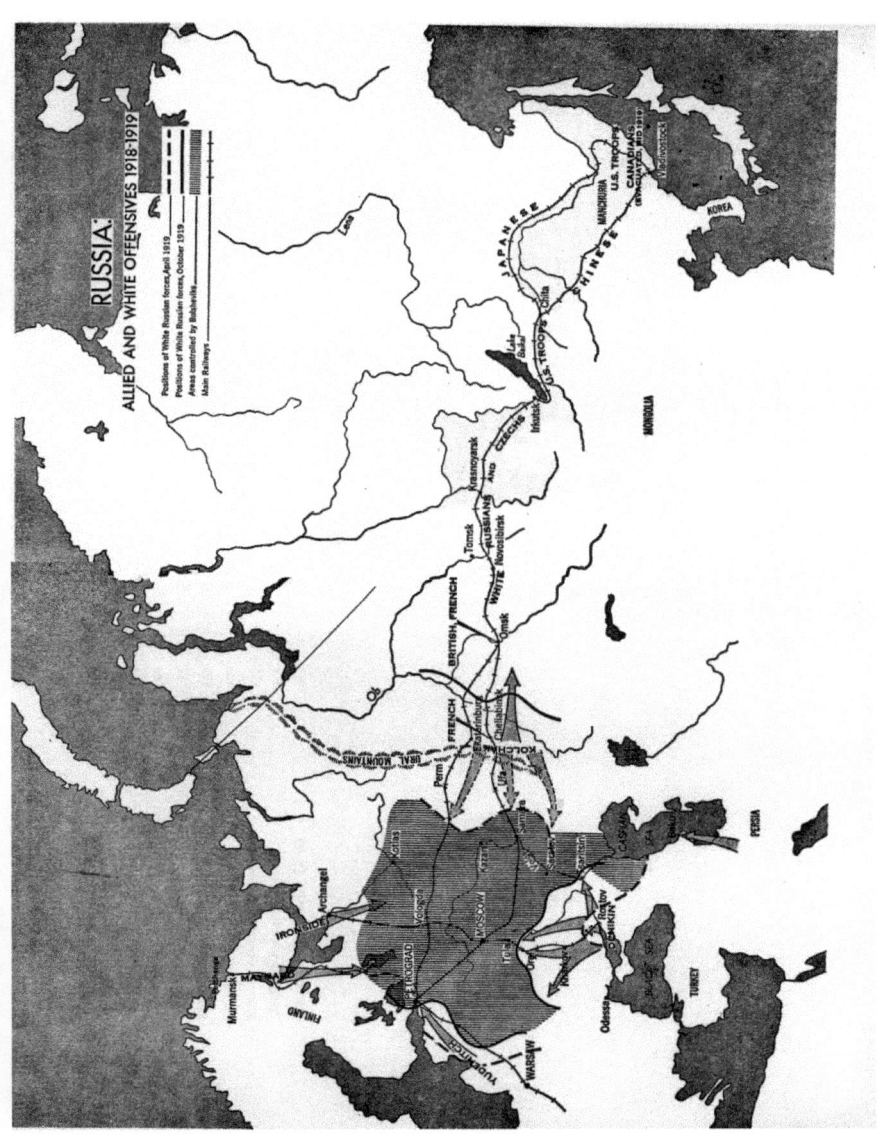

ARCHANGEL AND GALLIPOLI

"This war was one of the most unjust ever waged. It was an instance of a republic following the bad example of European monarchies."

**From Personal Memoirs of U. S. Grant.
Commenting on the war with Mexico.**

ARCHANGEL AND GALLIPOLI

"Theirs not to reason why;
Theirs but to do and die."

MANY people have asked me about the Russian campaign, why American soldiers went to Siberia, and what they did after they got there, for the general notion seems to be that Russia and Siberia are synonymous, and that the Russian Expedition, whatever its hazy purpose was, was centered about Vladivostok, and that in this far eastern port, a few American and Allied soldiers "marked time," while their comrades on the Western Front fought out, and eventually conquered, in the greatest of all wars. One American officer was actually ordered to join his command at Archangel, "via Vladivostok," and the order was issued by the War Department of the United States. Six thousand miles of inaccessible territory separated these two Russian ports, and the average American soldier who went out from Archangel in the fall of 1918, and, during the desolate winter months that followed, fought for his life along the Vologda railway, or far up the Dvina river, or in the snows of Pinega and Onega valleys, never knew that Brigadier General William S. Graves of the United States army, with thirteen hundred eighty-eight regulars and forty-three officers, had landed at Vladivostok on 4th September, 1918, and remained there after the Archangel fiasco had terminated. There was no conscious liaison between this American company of the far East and that of the far North, each performing burlesque antics in fantastic sideshows, while in the West, the greatest drama of all time was in its denouement, and a tense world trembled as it watched.

Whether there was any political connection between the Archangel Expedition and the Vladivostok Expedition is for the statesmen to answer. Surely there never was any military connection. Obviously, there never could be any support or communication between the two forces, and the American soldier at the Arctic Circle who was not told the reasons why he faced death and unknown dangers there, and why he was weakened and broken, and made old by privation and intense cold, never knew that there was a Siberian Expedition, and does not know even to this day.

So I have thought it worth while to tell, as faithfully as I could, the story of this strange war of North Russia, an insignificant flickering in the glare of the mighty world conflict, but inspiring in its human significance, its exploits of moral strength and sheer resolution and godlike courage. I have considered the campaign as a trial by ordeal of American manhood, that tested our souls to the depths, like Gallipoli tested the British. It was like Gallipoli in the hopeless odds encountered at every turn, in the vague outline of the commitment at the outset; in its distressing losses; its hardships and privations; its tragical ending.

But it was very vitally unlike Gallipoli, because in the war with Russia the soldier never knew why. The Australians, in their effort to force the Dardanelles, were exalted by the belief that theirs was an important operation in the war, and the British soldier went to battle the Turks, convinced that if he died, it was to save some little spot in a Cheshire or Sussex village, which to him meant England. It was a holy war, and men were fired with the high, selfless devotion of the Crusaders. An arrogant, brutal power swaggered abroad, menacing liberty, and the home and all things of the spirit. If German Imperialism engulfed civilization, there would be nothing left to live for anyway.

But there were no such reflections to sustain the soldier in Russia. The Armistice[1] came, and he remembers the day as one of sanguinary battle, when his dwindling numbers suffered further grievous losses, and he was sniped at, stormed with shrapnel and shaken by high explosive shells. He heard of the cessation of blood-letting in France and Belgium, but for many desolate, despairing months, he stood to his guns, witnessing his comrades killed and mutilated, the wounded lying in crude, dirty huts, makeshifts of dressing stations, then in sledges, dragged many excruciating miles over the snow to the rear, where often they got little better attention than at the front lines. He knew his physical strength was failing under the unrelieved monotony of the Arctic exploration ration; he saw others with scabies and disgusting diseases of malnutrition, and wondered how long before he too would be in the same way. He felt his sanity reeling in the short-lived, murky, winter days, the ever encircling menace of impending disaster and annihilation. He asked his officers why he fought, and why he was facing an enemy vastly superior to him in strength and equipment and armament, and why he was separated from his family and home and the ways of life, and when the end would come. But his officers were silent under this inquisition. They asked the same questions themselves,

1   November 11, 1918, was the day the First World War officially ended. At the same time, on that very day, American troops were in North Russia fighting for their lives.

and got no reply. The colonel who commanded this fated regiment told his soldiers that he could give no reason for them to oppose the enemy other than that their lives and those of the whole expedition depended upon successful resistance.

So soldierlike, he "carried on," while the dreary skies above him menaced death, and death stalked the encompassing forests of the scattered front lines, and the taint of death was in the air he breathed.

In the end, and when nearly all hope had fled, he returned homeward, stricken in health and dazed in spirit, where people moved as before, and were agitated by the same concerns, as if nothing had occurred to upset the whole scheme of things and uproot forever the old standards of values and ambition and morality. They noticed a queer look in his eyes and that he was customarily silent, often introspective. They manifested a casual interest in his great adventure. They never could understand.

Both expeditions were conceived by the British High Command and both were conducted by the execution of British military orders. Perhaps therein is the underlying philosophy of North Russia and Gallipoli; this attachment of the British mind to an astricted faith in England and her imperial destiny to rule the peoples of the world, contemptuous of obstacles and difficulties and perils in unknown, alien lands that appear very real to other than British mental processes.

"We'll just rush up there and re-establish the great Russian army—reorganize the vast forces of the Tsar," said an ebullient officer in England, wearing the red tabs and hatband of the General Staff. "One good Allied soldier can outfight twenty Bolsheviks," was the usual boast of the Commanding Officer in the early days of the fighting.

And it was a boast that was made good in the furious winter combats, when, standing at bay, the scattered companies, with no place to retreat, save the open snow, stood off many times their number of the enemy. In these decisive trials, the spirit of the Anglo-Saxon ever asserted its superiority, but one to twenty is not a very comfortable ratio upon which to form an offensive campaign. And the war against Russia was conceived as an offensive campaign, whatever it turned out to be.

Men of the 339th

RUSSIA THE VAST UNKNOWN

"The Emperor fully realized the nature of the task he had before him. To defend himself in Italy, Germany or even Poland against the Tsar was one thing; to invade the vast empire of Russia, was another task altogether—a task colossal, if not appalling. And arrayed against him were two fearful enemies — the Russian Army and winter."

WATSON'S Napoleon.

RUSSIA THE VAST UNKNOWN

SOMETIMES we are amused by foreign litterateurs and commentators, who come to our great country for a few crowded weeks of teas and symposia, gatherings of the intelligencia in our metropolis, and perhaps a dash into the mushroom dilettantism of Chicago, to set sail and compose screeds and screeds of America, her ways and her people, their manners and their customs.

Superficial vaporings, but far better composed and built by far on firmer ground than the idle opinions of those few Americans who have gone to the vast, far stretching empire of the Slavs, and glibly vouchsafed their ex cathedra views thereon.

The dominions of Great Russia were spread from the Baltic east to the Japan Sea, and from above the Arctic Circle far south to the Caspian and the Black Sea and Lake Baikal in Siberia. They comprised eight million six hundred and fifty thousand square miles of varied territory, nearly three times that of the United States, and were people by heterogeneous people, numbering one hundred and eighty million, as estimated, for no census or even approximate count has ever been attempted.

There were the Finns and the Letts, the Lithuanians, the Jews, the Mordvinians, the Estonians, the Siberians, the Great Russians, the Little Russians, the Red Russians, and the White Russians of the Central Provinces, the Cossacks of the south, and the Tartars of the Caucasus; all with no conscious unity, no national identity, not a single common impulse or purpose or interest. In many instances, without a communion of language.

The total length of railways in 1917 was thirty-four thousand miles, or less than one-eighth of that of our country. Of these one hundred and eighty million Russians, nearly eighty per cent are moujiks,[2] docile, patient serfs, liberated scarcely sixty years ago by Alexander II, and still shackled by the shackles of their serfdom, woeful ignorance, cowed spirit and afflicting poverty.

The remaining twenty per cent are survivors of the fading nobility and the bourgeoisie, or middle class, who have acquired wealth and consequent social rank without claim to nobility of birth. These last are hated with an intense, irrational hatred by the

---

2   i.e. muzhiks or **мужйк**, term for Russian peasant.

Bolsheviki.

The noble class, the Russian of Turgenev,[3] supersensitive, highstrung, supercultivated, almost to the point of degeneration, is fast vanishing with the passing of the last vestige of the Romanoff regime, and soon will be a thing of the past. This intolerant caste for centuries had dwelt in idleness on great landed estates. It was as alien to the poor moujik as if of an entirely distinct race. I met a few of these highborn on the streets of Archangel, whence they had fled from the murderous Reds in the cities of Moscow and Petrograd. Elegant gentlemen they were, in all the glittering panoply of Imperial army officers, and manners the extreme in politesse; very pompous, extremely impressive. They did not conceal their contempt of the crawling moujik; he was a swine, and when the word was hissed in Russian, it sounded very swinish.

The serf and the highborn, the swaggering, objectionable bourgeoisie, the moujik and his animal ignorance, the intelligencia, and his superculture, each separated from the other by an abysmal unspannable gulf; and the various Russian races so dissimilar in thought and living, in customs, even in language, all nevertheless were kept in some semblance of cohesion by the brutal, disciplinary methods of the Tsar and the cooperating spiritual guidance of the Russian State Catholic Church, of which the Tsar was the Little Father.

San Francisco is as acutely conscious of national affairs in Washington, as New York, and more so. But this is because the finest transportation system in the world makes it possible to journey from one city to the other, in a few days, and because every American is an ardent disciple of our great public press.

But Vladivostok knows nothing of Petrograd, and Petrograd knows little of Archangel, and in the little villages, where the people live, the world beyond is clothed in im-penetrable mystery; for there are no railways to these villages. No news comes in, and if news came, there are few among the moujiks who could read it.

It is well to keep these things in mind when men speak of Russia, as if overnight it could formulate a concerted policy and engage in a purpose backed by preponderant control of the Russian people. Russia is not a nation, it is an immense, unwieldy empire, a giant of tremendous strength, with undreamt-of potentialities, capable of colossal deeds, but without authoritative, united control or direc-

---

3   Ivan Turgenev (1818-1883) Russian aristocrat, poet, playwright, novelist.

tion; entirely unconscious of any national entity.

When Nicholas abdicated in March, 1917, it was an anxious world that viewed the experimental government of Prince Lvoff.[4] Russia was an important ally, but she had made heroic sacrifices and had lost five millions of men; if she faltered now, the world might be lost. And there were rumors of a separate peace.

A few months after the downfall of the Tsar, Kerensky, as Premier, issued a manifesto expressing undying allegiance to the sacred cause of the Allied Nations, and shortly delivered to the army his famous Prikaz,[5] which:

a. Abolished the penalty of death for disobedience of essential military discipline.

b. Abolished soldierly courtesy and the salute. Officers were henceforth to be known as *tvarishi*, comrades, and all social distinctions between them and the common soldier were abrogated.

c. Meetings of soldiers to discuss the conduct of military affairs were permitted.

Officers were simply unmanned of any effective authority. They were permitted to administer and instruct their organization, but all disciplinary measures were passed upon by a committee of soldiers, and so obedience to any order was a matter for ultimate ruling by such a soldier committee and not by an officer. This was democracy run riot, individual liberty gone stark mad. A few weeks after Kerensky took command, one million five hundred thousand Russian soldiers, grown weary of the tedium and the hazards of the front, quit the army and returned to their homes.

Thus by one foolhardy, ill-advised measure, an army became a rabble. Discipline, as essential to the military as blood is essential to sustain a physical body, vanished, and the collapse of Russia began with Kerensky.

After the entry of the United States into the war in April, 1917, President Wilson was uneasy about Russia and her future

4   Prince Georgy Lvov (1861-1925) came from one of the oldest noble families of Russia. He was appointed prime minister after Tsar Nicholas II abdicated from the throne. He took over in March 1917, and enacted wide reforms, including female suffrage. He could not maintain support, and resigned in favor of Kerensky. He left Russia in 1918, and never returned. He died in Paris in poverty, while writing his memoirs, in 1925.
5   Prikaz means 'order' or 'decree' or 'command.' He made these orders under heavy pressure from Lenin and the left.

course against the common enemy. Emissaries were therefore sent to learn of conditions first hand. Headed by the Honorable Elihu Root, as Ambassador Extraordinary, these reached Petrograd on the 13th June, 1917. Charles P. Crane, Cyrus H. McCormick of Illinois, and General Scott, the American Chief of Staff, accompanied Mr. Root. The emissaries met Kerensky, talked with several military and labor leaders, attended many banquets, made as many good speeches, and reported to the President in Washington on 12th August of the same year.

This report was made in confidence to the President, and even at the late date of the present writing, all requests to examine it have been denied by the State Department, on the grounds that "Divulgence is incompatible with the public interests."

But shortly afterwards, Mr. Root gave out an interview, which purported to express the views of the delegation: that they had come back with faith in Russia; faith in Russian qualities of character that are essential tests of competency and self government; faith in the purpose, the persistence and the power of the Russian people to keep themselves free.

Many American bankers, believing in Mr. Root, manifested kindred faith by the exchange of good American dollars for Russian rubles, despite the fact that the Russian government was hopelessly bankrupt and was showing an operating deficit of milliards of rubles.

General Scott visited the Russian front and witnessed the offensive which resulted in the taking of Kovel and Lemberg.[6] He conferred with Generals Brussiloff, Korniloff, and Erdeli[7] and their staffs, and reported to the American Secretary of War that Russia would stay in the war "if given even a part of the aid she asks."

Three months before the debacle, the Secretary of State, Mr. Lansing, assured the American people that Russia was stronger than she had been for some time, both from the government point of view and the military point of view.

The government point of view? The outstanding feature of the Russian Government "point of view" has always been the venal

---

6   i.e. the Brussiloff offensive, Брусиловский прорыв, or Brussiloff's breakthrough, a 1916 counter-offensive against Austria Hungary. The success galvanized Russian morale, and brought Romania into the war.

7   General Brussiloff was a Russian noble who would switch sides and serve the Red Army. Kornilov fought for the whites, and was killed in 1918. Erdeli fought for the Whites and later fled to France.

disposition of the High Command; the shameful, heartless, conscienceless corruption of persons in authority. Everyone knew this who knew Imperial Russia. At the trial of General Sukhomlinov,[8] Minister for War, General Yanushkevitch,[9] former Chief of Russian General Staff, testified that in the retreat from Galicia, during the summer of 1915, there was only one rifle for every ten soldiers. The soldiers in the rear had to wait until their comrades on the firing line were killed so that they might have their rifles. The Russians had no shells, and the Germans knowing this, set their guns two thousand yards off and shot down one helpless regiment after the other.

Many other examples of pitiful defenselessness could be cited at a time when the Allies loaned hundreds of millions of dollars to Russia for arms and military equipment, and Russia had these munitions, but far back of the front lines.

We have viewed Russian affairs as we have viewed Mexico, with American provincial eyes instead of attempting to judge from a Russian angle. Gladstone said that a nation guided by provincial statesmen was doomed for perdition, and, by reason of our provincialism, American statecraft striving to cope with Russia was hopelessly handicapped at the outset. This wholesale scandal and shameless corruption in high circles was typically Russian, an essential premise upon which to form a judgment of the Russian situation, but a premise totally unknown to persons unfamiliar with Russian character and Russian conditions.

Democracy assumes intelligence, but most important of all, self-control. Had we been familiar with the Russian people, is it likely that our State Department would have given such unstinted confidence to the dreamer, Kerensky? For like all countries where ignorance stifles the progress of struggling national life a strong unhesitant hand was needed to guide the nascent Russian democracy, and instead of resolution Kerensky presented oratory and by his Prikaz and vacillating policies rapidly lost his grip upon the army. General Korniloff attempted to rally the demoralized forces, restored the death penalty and strove to bring out of the chaos created by Kerensky, some likeness of coordination, but there was

8   He was arrested and tried for corruption in 1916. His much younger wife had expensive tastes. However, he seems to have been a scapegoat for other issues within the Russian Army. He left Russia, published his memoirs, and ended up in Berlin, in extreme poverty, found frozen to death on a park bench in 1926.
9   Yanushkevich was forcibly retired after the February Revolution. He was arrested and killed in 1918.

a division in adherence to the Premier and the General, and in the end both Korniloff and Kerensky failed. Probably no man could have succeeded; the seeds of destruction had germinated and struck root. It was too late.

The revolution of the Bolsheviks took place on 7th November, 1917, and in February following was announced the Peace Treaty of Brest-Litovsk, whereby the provinces of Russian Poland, Courland, Lithuania, and Estonia came under German control, giving Germany an important Baltic littoral. Turkey, the ally of Germany, was to receive back all territory in Asia Minor occupied since the war, and in addition the districts of Kars and Erivan and Batum. Germany and Turkey controlled the Caucasus, the boundaries of which were to be restored as they existed before the Russian-Turkish War of 1877. During the civil war that followed in the Ukraine, the Germans occupied the port of Sevastopol, and the Austrians took Odessa. Germany got vast stores of guns and war material, thirteen thousand three hundred fifty miles of railway, more than one-third of the entire Russian rail system, a large amount of rolling stock, seventy-three per cent of Russian iron fields and eighty-nine per cent of her coal.

The war in the East was over, one hundred and forty-seven German and Austrian Divisions were released for the Western Front.

Archangel, where the East comes abruptly face to face with the West

American Soldiers in front of an Archangel Church

OBJECTS OF THE EXPEDITION

"Shall the military power of any nation or group of nations be suffered to determine the fortunes of peoples over whom they have no right to rule except the right of force?"

WOODROW WILSON — 27th September, 1918.

OBJECTS OF THE EXPEDITION

IT IS said of the Bolsheviks, that they are a terrorist, minority party, rode to power by the seizure of every available machine gun in Russia and maintain their sway by the same forceful persuasion.

One of the *intelligencia* once told me, that of every hundred Russians, only two were Bolsheviks, and the remaining ninety-eight were cowed into submission by the methods of the desperado.

This, to enlightened, high-spirited America is a preposterous statement, but Russia is not America. Nor has she America's schools, nor America's great railways, nor the public press of America.

At Brest-Litovsk, Russia was stripped of nearly all war supply and munitions by the unsparing Germans, and what was left was seized by the belligerent Soviets.

Now, even in proud America, a resolute man back of a six shooter has been known to hold up an entire train load of people. And whether the Soviets are backed by the sanction of the masses, or whether as the Imperialists would have us believe, they are an unprincipled, bullying minority, they are in truth and fact the de facto government and represent the sovereignty of Russia in the comity of nations.

For six years Lenin and Trotsky have ruled,[10] while the ministries of America, France, England and Italy have undergone complete transformation with the changing judgments of these troublous times, and now, begrudgingly, Russia; Russia of the Soviet Party, proletarian Russia, anarchistic, "nihilistic" Russia is given a seat at the international conference table of Lausanne, Great Britain has officially recognized the Soviets, and clamorous politicians in this country (even one statesman), are emphatically demanding recognition by the United States.

The Bolsheviks derived their inspiration from the Russian anarchist, Bakunin, an apostle of terror and violence. Bolshevik comes from the Russian word *bolshinstvo*, the majority. The name was used for the first time in 1903, when Nicolai Lenin split the Social Democratic party in two and assumed leadership of the majority.

---

10   Cudahy was writing in 1923. Lenin would die in 1924. Trotsky was undermined and isolated after 1924, and expelled from the Soviet Union in 1929.

Lenin's real name was Zederblum, that of Trotsky, Bronstein.[11]

The moving purpose of Bolshevism is to organize a great inter-national revolution, affecting all countries. A revolution that will eradicate forever the hated capitalist class, and the despised small proprietors and entrepreneurs, known as bourgeoisie. Bolshevism is openly an enemy of democracy. It has no tolerance for any class save the proletarian. In the Bolshevik era, only the proletariat has any claim. Bolshevism is autocracy, autocracy of the proletariat. A ruthless autocracy that would utterly destroy every social group except this favored one.

Directly after he assumed power, Lenin put into effect the Land Decree, which abolished the title of landlords to real estate and confiscated all landed estates, except the small holdings of the peasants. All employers of labor were suppressed, the six-hour day was established in industrial enterprises, and all employees were to have a voice in the management.

There is naught in this program which can be reconciled with German Imperialism, yet many statesmen and soldiers in Allied councils were convinced that an alliance existed between the Bol-sheviks and Germany. But it is impossible to conceive of two more extreme opponents in political philosophy, for the Prussian Junkers believed devoutly in the divine commission of kings, as enunciated by the Kaiser himself; and the Bolsheviks, hating every suggestion of imperialism with an intense, raging hatred, threatened death to every king, and recognized, as qualified to rule or govern, none save the proletariat.

Only one tenet did Bolshevism and Prussian militarism have in common, i.e., they were both invincibly opposed to democracy. Both archenemies of political justice, as we Americans understand political justice.

The military leaders and statesmen at Berlin beheld with serious alarm the Revolution of November, 1917. They loathed the Bolsheviks and feared the effect of their insidious propaganda on the German masses. The German Chancellor, Von Bethmann, was obsessed with the fear of Bolshevism, and Ludendorff writes bitterly of the grave error in failing to crush the Soviet Party and

---

11   This is a can of worms we cannot answer in a footnote. The 'Zederblum' alias can be found in other newspapers and books from the time, and seems linked to his time in Switzerland. This fact was covered up under Stalin and for the rest of the cold war. Declassified documents, including a letter from Lenin's sister, suggest that he had Jewish ancestry via Moses Blank, his grandfather.

to openly take sides with its opponents in Russia. He speaks of the lowered morale of the Eastern German Divisions; how several of them proved utterly worthless in the battles of France, as a consequence of coming in contact with the Bolsheviks;[12] how the Bolshevik revolutionary ideas corroded the spirit of the people at home, and had more to do, than the military defeat, with the downfall of the German Government.

And the Soviet leaders returned the venom of Berlin with even greater virulency. They denounced the Brest-Litovsk agreement, stigmatizing it as: "The rape of Russia," and in their propaganda repeatedly expressed imperishable hatred of the German Imperialists. Lenin withdrew from the negotiations at Brest-Litovsk on 11th February, 1918, and refused to accede to the harsh demands of Germany. Thereupon, the Ukraine was immediately invaded, and on 1st March, the Germans occupied Kiev, the capital, holding a line to Reval on the Gulf of Finland, through Estonia, Pskov, Vilebsk and Mogilev. The helpless Russians could do nothing but submit, and under duress signed the treaty on 3rd March, 1918.

Still has it been affirmed by Allied statesmen time and repeatedly that the Bolsheviks were a willing party to the Brest-Litovsk pact, and that Moscow and Berlin were conspiring for the destruction of all Western civilization.

In his Fourteen Point address to Congress on 8th January, 1918, President Wilson expressed deep sympathy with Russia and enunciated Point VI as one of the cardinal principles for which the Allies fought:

> *VI. The evacuation of all Russian territory and such a settlement of all questions affecting Russia as will secure the best and freest cooperation of the other nations of the world in obtaining for her an unhampered and unembarrassed opportunity for the independent determination of her own political development and national policy, and assure her of a sincere welcome into the society of free nations under institutions of her own choosing; and, more than a welcome, assistance also of every kind that she may need and may herself desire. The treatment accorded Russia by her sister nations in the months to come will be the acid test of their good will, of their comprehension of her needs as distinguished from their own interests.*

---

12   It was Bolshevik policy to spread as much propaganda to the German Army as possible, to fraternize with them as much as possible, and to encourage them to set up their own Soldier's 'Soviets' or Councils, with the aim of encouraging a 'world revolution' instead of 'socialism in one country.'

On 11th March, 1918, on the eve of its meeting to pass upon the question of the acceptance or rejection of the Brest-Litovsk terms, the President sent a message of friendship to the all Russian Congress of Soviets, which contained this pledge:

> *Although the government of the United States is unhappily not now in a position to render the direct and effective aid it would wish to render, I beg to assure the people of Russia, through the Congress, that it will avail itself of every opportunity to secure for Russia once more complete sovereignty and independence in her own affairs and full restoration in her great role in the life of Europe and the modern world. The whole heart of the people of the United States is with the people of Russia in the attempt to free themselves forever from autocratic government and become masters of their own life.*

Many contend that if the Allies had stood by the de facto government of Russia, as President Wilson's words gave promise of doing, the disastrous treaty would never have been accepted.

Questions have been addressed to the then American Secretary of State asking: Did the administration know at the time of the Brest-Litovsk negotiations:

1. That the Soviet government represented by Lenin and Trotsky was opposed to the projected treaty and signed it only because of the physical impossibility of resisting German demands unless some of the Allies came to its aid?

2. That Lenin and Trotsky gave a note to Colonel Raymond Robbins of the Red Cross, stating to the President of the United States that they were opposed to the treaty and would not sign if the United States would give food and arms to the Russians?

The reply of Mr. Lansing was that answers to these questions were not compatible with the public interest.

On 12th December, 1918, Senator Johnson asked this question in the United States Senate:

> *Is it true that the British High Commissioner, sent to Russia after the Bolsheviki revolution because of his knowledge and experience in the Russian situation, after four months in Russia, stated over his signature that the Soviet government had cooperated in aiding the Allies, and that he believed that intervention in cooperation with the*

Soviet government was feasible as late as the fifth of May, 1918?

No spokesman for the administration, or anyone else, ever answered or attempted to answer this question.

After Brest-Litovsk, it was generally believed that the ambitions of Germany in Russia were:

1. To recruit her war wasted divisions from the great number of Austrian and German prisoners in Russia.

2. To exploit the great natural resources of the Ukraine, Courland, Lithuania and Estonia.

3. To align on her eastern frontier buffer states from Finland to the Caucasus with Persia as the last link in the chain.

4. To seize great stores of war munitions at Archangel and Vladivostok.

There was also some credence in the rumor that Germany sought to establish submarine bases at Murmansk and Petchenga in Finland.

Murmansk, on the Kola Peninsula, is the only port of North Russia not closed for nearly half the year. During the months of winter, from December until the middle of June, Archangel, Kem, Onega and Kandalaksh on the White Sea are sealed by effective barriers of ice, and even Petrograd, several hundred miles further south on the Baltic, is closed until late in April. But the Cape current of the Gulf Stream swings around the northern coast of the Kola Peninsula, and at Murmansk there is an excellent natural harbor, which is always open, with thirty-two feet of water in shore, and a high coast line, giving splendid protection against storm. From this valuable ice free port, the Murman railway extends three hundred miles to Kem and continues through Petrozavodsk on the west shore of Lake Onega, six hundred miles further to Petrograd.

The completion of this, the most northern railroad, is a triumph of imagination and courage and invincible resolution. The Russian engineer, Goriatchkovshy, inspired by the necessity of his country having a means of inlet for munitions and supplies during the war (for the Trans-Siberian railway could carry only about one-seventh of such supplies), laid the tracks over seemingly bottomless tundra and conquered in the face of most disheartening discouragement.

A great number of German prisoners and one hundred thousand Russian laborers worked to complete the heroic enterprise.

Experts predicted that with the melting of the ice in spring, the tracks would disappear in the marshes, but Goriatchkovshy had reckoned with the elements. The Murman railway is operating today. It has a hauling capacity of thirty-five hundred tons a day, the maximum handling facilities of Murmansk port, and many a lonely soldier, snowbound in North Russia, during the tragic winter of 1919, has the Murman railway and its creator, Goriatchkovshy, to thank for the messages from far off America, that came to Murmansk and were brought to Archangel by Obozerskaya on the Vologda railway, and then relayed by droshky and the faithful Russian pony to a solitary sentinel post somewhere in the great white reaches of the interior.

Very close to the Murman road is Finland, which, because of its remoteness from the Russian capital, had always exercised a limited autonomy, and following the Kerensky Revolution of March, 1917, announced by the action of the Finnish Diet, its complete independence.

A civil war between Red Guards and White Guards for the control of the government followed. It was no secret that from the beginning of the European war the sympathies of the Finns were with Germany, and now at the outbreak of this internal conflict in Finland, Germany aligned with the White Guards against the revolutionary Reds who were supported by the Bolsheviks.

At the beginning of April, 1918, three regiments of German rifles, two batteries and three battalions of Jägers, under General von der Goltz, landed at Hanko, and, cooperating with the White Finns, suppressed the revolution, took possession of the port Viborg and were in control of railway communication to Petrograd. But this small expeditionary force never left the southern part of Finland, and in August, when every German was needed in France, the greater part of it left for the Western Front.

The campaign in Finland had no effect on the course of the war. Its significance was unduly magnified by both sides.

It was a firm conviction in Allied Councils that the Germans had immense forces in Finland, while the German Imperial Staff thought that the insignificant hundreds that the British landed at Murmansk in April, almost at the same time that the Germans entered the south of Finland, were in large numbers, perhaps several Divisions.

Thus there existed a blindman's buff in Finland; both Commands in startling ignorance of enemy salient facts, which is often the case in the game of war where "uncertainty is the essence"; each supposed the other was actively engaged in "recreating an Eastern Front," which, in concrete application, meant the recruiting of hundreds of thousands of Russians to press on from the East and fill in the war-wasted gaping ranks of Germany or the Allies.

To effect this object and gain access to the interior of Russia, the Murman railway, therefore, assumed a momentous significance; but in truth the "Eastern Front" remained a figment of the military imagination. Russia had poured out the life blood of her sons in the Allied defense till she staggered weak and exhausted, so spent that she swayed in a moral lethargy from which nothing on earth could arouse her, and those Russian soldiers who survived returned to their villages or else were conscripted for the Red army by the amazingly effective methods of Trotsky.

Still, in the spring of the year 1918, the situation in Finland appeared so fraught with grave potentialities of decisive consequence, that on 27th May, the Allied military attaches of Italy, France, England and the United States met at Moscow and unanimously agreed that these nations should intervene in the affairs of Russia.

Shortly after this, the Supreme War Council at Versailles decided in favor of intervention in the northern Russian ports, and the United States gave its consent.

Brigadier General F. C. Poole had been in Petrograd in command of the technical war mission of the British in Russia. Thoroughly familiar with Russian character and Russian conditions, he was chosen to command the Northern Expedition.

The advance party of the Americans landed in Archangel on 3rd August, 1918. On the same day, this statement was cabled to the Russian Ambassador from the State Department at Washington:

> *In the judgment of the government of the United States, a judgment arrived at after repeated and very searching considerations of the whole situation, military intervention in Russia would be more likely to add to the present sad confusion there than to cure it, and would confuse rather than help her out of her distresses, as the government of the United States sees the present circumstances, therefore military action is admissible in Russia*

now only to render such protection and help as is possible to the
Czecho-Slovaks against the armed Austrian and German prisoners
who are attacking them, and to steady any efforts at self-govern-
ment or self-defense in which Russians themselves may be willing
to accept assistance. Whether from Vladivostok or from Murmansk
and Archangel, the only present object for which American troops
will be employed will be to guard military stores which may be
subsequently needed by Russian forces, and to render such aid as
may be acceptable to the Russians in the organization of their own
defense.

The importance of guarding the Arctic ports from the Germans
passed with the signing of the Armistice, but armed intervention
continued, and the most sanguinary battles in North Russia were
fought in the dark winter months that followed.

When the last battalion set sail from Archangel, not a soldier
knew, no, not even vaguely, why he had fought or why he was going
now, and why his comrades were left behind—so many of them
beneath the wooden crosses. The little churchyards and the white
churches and the whiter snow! Life will always be a crazy thing to
the soldier of North Russia; the color and the taste of living have
gone from the soldier of North Russia; and the glory of youth has
forever gone from him.

It is a fearful thing to contemplate the deliberate taking of a
life. All consciousness recoils at the dreadful, irretrievable consequences of murder; yet when nations engage in extensive killing,
there is no malice in the act on the part of individuals. Killing then
has an impersonal character and becomes an heroic contemplation.

In Western trenches, the enemy was called "Jerry" in a spirit
of grotesque comradery and sportsmanship, and the finest soldiers
had little hatred in their hearts for those across the twisted, shell
gashed acres, who sought to maim and kill them, but with no malice aforethought.

The mildest men, and men of highest culture and intelligence,
recently made a profession of killing, and could practice their
newly found profession with keen, cold, ghoulish precision and the
comprehensive analysis of trained minds. War is not murder, and
the business of killing loses its infamy and much of its obscenity by
the united impulse of millions striving with selfless purpose, pure
devotion and heroic sacrifice for a nation's goal. War shears from
a people much that is gross in nature, as the merciless test of war

exposes naked, virtues and weaknesses alike. But the American war with Russia had no idealism. It was not a war at all. It was a freebooter's excursion, depraved and lawless. A felonious undertaking, for it had not the sanction of the American people.

During the winter of 1919, American soldiers, in the uniform of their country, killed Russians and were killed by Russians, yet the Congress of the United States never declared war upon Russia. Our war was with Germany, but no German prisoners were ever taken in this lawless conflict of North Russia, nor, among the bodies of the enemy killed, was there ever found any evidence that Germans fought in their ranks or sat, in the councils of their Command. And in the conduct of the whole campaign there was no visible sign of connection between the Bolsheviks and the Central Powers.

The war was with the Bolsheviki, the existing government of Russia, and a few weeks after the arrival of American troops in Archangel, Tchitcherine, Soviet Commissioner for Foreign Affairs, handed a note to Mr. Christiansen, Norwegian diplomatic attache, which was delivered to President Wilson, in which the Bolsheviks offered to conclude an armistice upon the removal of American troops from Murmansk, Archangel and Siberia.

This note was ignored. The Soviets had no recognition as the government of Russia, and there was no "war" in Archangel or Murmansk or Siberia.

No war, but in the province of Archangel, on six scattered battlefronts, American soldiers, under British command, were "standing to" behind snow trenches and improvised barricades, while soldiers of the Soviet cause crashed Pom Pom projectiles at them, and shook them with high explosive and shrapnel, blasted them with machine guns, and sniped at any reckless head that showed from cover.

The objects of the Expedition, as defined in a pamphlet of information given out by British General Headquarters in the early days of the campaign, were:

1. To form a military barrier inside which the Russians could reorganize themselves to drive out the German invader.

2. To assist the Russians to reorganize their army by instruction, supervision and example on more reasonable principles than the old regime autocratic discipline.

3. To reorganize the food supplies, making up the deficien-
cies from Allied countries. To obtain for export the surplus
supplies of goods, such as flax, timber, etc. To fill store ships
bringing food, "thus maintaining the economical shipping
policy."

The Bolshevik government is entirely in the hands of the
Germans, who have backed this party against all others in Russia
owing to the simplicity of maintaining anarchy in a totally disorganized country. Therefore, we are definitely opposed to the Bol-
shevik-cum-German party. In regard to other parties, we express
no criticism and will accept them as we find them, provided they
are for Russia, and therefore "out for the Boche." Briefly, we do
not meddle in internal affairs. It must be realized that we are not
invaders, but guests, and that we have not any intention of attempting to occupy any Russian territory.

Later, this proclamation was issued to the troops by the military
authorities:

> *Proclamation: There seems to be among the troops a very indis-
> tinct idea of what we are fighting for here in North Russia. This
> can be explained in a few words. We are up against Bolshevism,
> which means anarchy pure and simple. Look at Russia at the present
> moment. The power is in the hands of a few men, mostly Jews, who
> have succeeded in bringing the country to such a state that order is
> non-existent. Bolshevism has grown upon the uneducated masses to
> such an extent that Russia is disintegrated and helpless, and there-
> fore we have come to help her get rid of the disease that is eating her
> up. We are not here to conquer Russia, but we want to help her and
> see her a great power. When order is restored here, we shall clear out,
> but only when we have attained our object, and that is the resto-
> ration of Russia.*

At about the same time that this proclamation was spread
among British soldiers in Russia, the Inter-Allied Labor Conference met in London and sent an expression "of deepest sympathy
to the labor and socialist organizations of Russia, which having
destroyed their own imperialism, continued an unremitting struggle against German Imperialism."

Still later, there was broadcasted among the soldiers, headed
"Honour Forbids," an exposition of the campaign by Lord Milner,
British Secretary of State for War, who defined its objects:

1. To save the Czecho-Slovaks. Several thousand of which under command of General Gaida were believed to be strung along the Siberian railway from Pensa to Vladivostok.

2. To prevent the Germans from exploiting the resources of Southeastern Russia.

3. To prevent the northern ports of European Russia from becoming bases for German submarines.

When these objects were accomplished, the British statesman declared that to leave Russia to the unspeakable horrors of the Bolshevik rule would be an abominable betrayal of that country, and contrary to every British instinct of honor and humanity.

During the winter months of 1919, when Senator Johnson[13] was demanding in the United States Senate the reasons for the American war with Russia Senator Swanson,[14] of Virginia, of the Foreign Relations Committee, and one of the spokesmen of the administration replied that American troops were needed to protect great stores of Allied ammunition at Archangel, and to hold the port until terms of peace were signed with Germany. That Germany wanted Archangel to establish a submarine base there, and it would be cowardly to forsake Russia.

During the peace negotiations at a meeting of the Council of Ten at Quai D'Orsay, on 21st January, 1919, President Wilson, in discussing the Russian problem, stated that by opposing Bolshevism with arms the Allies were serving the cause of Bolshevism, making it possible for the Bolsheviks to argue that imperialistic, capitalistic governments were seeking to give the land back to the landlords and favor the ends of the monarchists. The allegation that the Allies were against the people and wanted to control their affairs provided the argument which enabled them to raise armies.[15] If, on the other hand, the Allies could swallow their pride and the natural repulsion which they felt for the Bolsheviks, and see the representatives of all organized groups in one place, the President

13 Senator Hiram Johnson of Indiana, (1866-1945) gave an outspoken speech against this undeclared war in Russia, and we have reproduced it in appendix below. He served as Governor of California from 1911-1917, and Senator from California from 1917-1945. He was a Progressive Republican aligned with Teddy Roosevelt, and ran in 1912 as Teddy Roosevelt's Vice President.
14 Claude Swanson (1862-1939) was a lawyer from Virginia, and a staunch Southern Democrat. He served as a Congressman from 1893-1906, Governor of Virginia from 1906-1910, Senator from 1910-1933, and Secretary of the Navy from 1933-1939.
15 Over 15 countries deployed troops into Russia during the Civil War.

thought it would bring about a marked reaction against Bolshevism.

Mr. Lloyd George, earlier in the discussion, said that *the mere idea of crushing Bolshevism by a military force was pure madness*. Even admitting that it could be done, who would occupy Russia? If he proposed to send a thousand British troops to Russia for that purpose, the armies would mutiny.

It was agreed by the Council of Ten, then Four, that President Wilson should draft a proclamation inviting all organized parties in Russia to attend a meeting in order to discuss with the representatives of the Allied and Associated Great Powers the means of restoring order and peace in Russia. Participation should be conditional on a cessation of hostilities. This meeting was to take place on Prinkipos Island in the Sea of Marmora.

The President issued the proclamation, but the French were opposed to it and communicated with the Ukrainians and the other anti-Soviet groups in Russia, to whom, as well as to the Bolsheviks, the proposal was addressed, telling them that if they refused to consider the proposal, the French would support them and continue to support them, and not allow the Allies, if they could prevent it, to make peace with the Russian Soviet government. The time set for the gathering at Prinkipos was on 15th February, 1919, but no party acted in a definite way and it never took place.

At the time of the Bolshevik revolution, the national debt of Russia was 700,000,000,000 of rubles. The interest and sinking fund charge was 4,000,000,000 of rubles annually. There was a deficit in the annual budget of one milliard. Of this total debt, 15,500,000,000 of rubles were owing to France, and France felt the prospective loss far more than any of the other creditor nations, for the French government had encouraged the purchase of rubles by her nationals, and these now nearly worthless securities were held by the peasants from Artois to Gascony.

Like the Prinkipos proposal, nothing came of a Soviet proposal for peace which was brought to the Paris Peace Conference by an emissary dispatched by the American commissioners to obtain from the Bolsheviks a statement of the terms upon which they were ready to stop fighting. This was in February, after the desperate situation of the troops near Archangel was brought to the attention of the Conference by the Allied Military commanders. These Soviet peace terms were approved by Colonel House at Paris,

who referred them to the President, "but the President said he had
a one track mind and was occupied with Germany at the time, and
could not think about Russia, and that he left the Russian matter
all to Colonel House."[16]

The sessions at Versailles adjourned without day. If we were
at war with Russia in 1919, we are still at war with her. Peace was
never made with Russia; and peace never will be made in the hearts
of those plain people in the Vaga and Dvina villages, who saw their
pitifully meager possessions confiscated in the cause of "friendly
intervention," their lowly homes set ablaze and themselves turned
adrift to find shelter in the cheerless snows.

Friendly intervention? All too vividly comes to mind a picture
during the Allied occupation of Archangel Province while the
statesmen at Paris pondered and deliberated in a futile attempt to
find dignified escapement from this shameful illegitimate little war.
Military necessity demanded that another village far up the Dvina
be destroyed. As the soldiers, with no keen appetite for the heartless job, cast the peasants out of the hones where they had lived
their uncouth, but not unhappy lives, the torch was set to their
houses, and the first snow floated down from a dark, foreboding
sky, dread announcer of the cruel Arctic winter. Within these crude,
log walls, now flaming fire, had they lived, these gentle folk, as their
fathers had lived before them, simple, unsophisticated lives, felici-
tously unmindful of petty vanities and corroding ambitions. Who
can say theirs was not the course of profoundest wisdom? For had
they not known in these humble homes those candid pleasures, the
only genuine ones, those elemental joys, springing like hope and
the unreasoning urge of life from the heart of humanity, oblivious
of all artificial environment? Here in these mean abodes had they
tasted the ecstasy of love, known the full poignancy of sorrow, wept
in natural grief and laughed loud with boisterous, unrestrained,
rustic laughter. In a corner hung the little ikon, where the lamp
burned on holidays, and they worshipped their God with a devotion so genuine, so deep and reverent, that only a fool could scoff.

Outside now, some of the women ran about, aimlessly, like
stampeded sheep; others sat upon hand fashioned crates, wherein
they had hastily flung their most cherished treasures, and aban-

---

16   Edward House (1858-1938), known as Colonel House, an honorary
title he was appointed by Texas Governor Hogg. He was a Texas businessman and
eminence grise who played kingmaker in Texas politics. He later became President
Wilson's right hand man and was his lead negotiator for the Treaty of Versailles.

doned themselves to a paroxysm of weeping despair; while the children shrieked stridently, victims of all the visionary horrors that only childhood can conjure. Most of the men looked on in spellbound silence, with a dumb, wounded look in their eyes. Poor moujiks! They did not understand, but they made no complaint. *Nitchevoo*,[17] fate had decreed that they should suffer this burden.

Why had we come and why did we remain, invading Russia and destroying Russian homes? The American consul at Archangel sent us the Thanksgiving Day message of our President, rejoicing in the Armistice, and the end of the carnage of war. But the consul announced that we would remain steadfast to our task until the end. The end! What was the end?

The British General Finlayson of Dvina Force said: "There will be no faltering in our purpose to remove the stain of Bolshevism from Russia and civilization." Was this, then, our purpose through the dismal night of winter time, when we burned Russian homes and shot Russian people? And was this still our purpose when we quit in June with Bolshevism strengthened by our coming, and more than ever before the government of Russia?

The only stain was the stain of dishonor we left in our retreating path. But a deep, red, burning stain of shame is on the foreheads of those men who sit on cushioned seats in the high places, chart armed alliances in obscure inter-national commitments, and, with careless gesture of their cigar, send other men to some remote forsaken quarter of earth, where there is misery and suffering, and hope dies, and the heart withers in cold, black days.

Now it was of small concern to Ivan whether the Allies or the Bolsheviks won this strange war of North Russia. What he heard was some vagary of "friendly intervention"; of bringing peace and order to his distracted country. What he saw was his village a torn battle ground of two contending armies, while the one that forced itself upon him, requisitioned his shaggy pony, took whatever it pleased to take, and burned the roof over his head.

He asked so little of life, this gentle moujik, with his boots and his shabby tunic, and his mild, bearded face, only to be left alone. In peace to follow his quiet ways, an unhurrying, unworrying disciple of the philosophy of *nitchevoo*.

---

17   Ничего, i.e. nothing, nevermind, no matter, no problem. A catchphrase conveying Russian fatalism.

28597   Sailors from the U.S.S. Olympia, who formed a part of a landing
force, returning from the line along the railroad to Vologda where
they had been fighting Bolsheviks.   The party got back to its
starting point only after picking its way through swamps and forests.
They were compelled to abandon everything but their rifles.   They
are surrounded by men of the 339th Infantry, who had just landed in
Russia.   Bakharitza, Russia.   Sept.6,1918.

THE PLAN OF CAMPAIGN

"I consider it my duty to inform you in plain language that unless considerable reinforcements are sent before the end of October, the military situation both at Archangel and the Murman Peninsula will, in my opinion, become very serious."

ADMIRAL KEMP, in command of British warships at Murmansk; to the Admiralty, 26th August, 1918.

61113   American soldiers on patrol wearing white capes to reduce the chance of discovery while operating in the snow-blanketed forrests which line the Vologda R.R. line on each side. The picture shows the character of the forrest in which the American forces live and fight. Left to right Bugler Chas. Metcalf. Co. I. Prt. Harold Holliday, Co. M. and Sgt. Maj. Ernest Reed, 3rd Bn. 339th Inf. 85" Div. Verst 455 Vologda R.R. Front, Russia. Feb. 21, 1919.

THE PLAN OF CAMPAIGN

THE Province of Archangel stretches from the Norwegian frontier across the Arctic Ocean east of the—Ural Mountains of Siberia. It includes the Kola Peninsula, which lies well north of the Arctic Circle, and the furthermost point south is below sixty-two degrees latitude. The total area is six times that of the average American state.

It is a poverty distressed and cheerless, destitute region;— which, during the reign of the Romanoffs, like Siberia, was often a place of exile and asylum for political dissidents. War accentuated the poverty of the province, and the only remanent sign of former industry is at the port of Archangel, where large timber mills, owned mostly by British capital, line both sides of the harbor.

The port was founded by Ivan the Terrible during the Sixteenth Century, and ever since then has been a British trading post. At Onega, Kem and Kamdalaksh on the White Sea, there is, or was, before the war, some small traffic in timber products, furs and flax. But this commerce is of small consequence. Prenatally, Archangel was destined for pauperism, for it lies in the far north, where life is poor and hard struggling, and there is little soft sunshine to woo riches from the earth. Nor are treasures concealed beneath its sear and barren surface. The curse of sterility taints the air, and it was never written in the Divine Plan that man should dwell in this fortuneless, forsaken region. He was banished there, or driven by the pitiless pursuit of his own misdeeds. For nearly half of the year, the White Sea is an impenetrable ice barrier, and then communication with the world beyond can be had only through the Murman railway to the far north port of Murmansk.

In the city, the East comes abruptly face to face with the West. The exotic colors of the great domed cathedral were brought from ancient Byzantium, when the Greek church was made the faith of his country by Vladamir; and bearded, sad-faced priests, with their black robes, glide through the streets like nether spirits, and the mysticism of the ancient, mystic East.

This is the native atmosphere of Archangel, and it will not be in a generation that the city will, without consciousness, take on the soft adornments and the practical utilities of Occidental civilization. The glaring electric lights, the incongruous, modern buildings and the noisy tramway that clangs down the street—these do not belong to Archangel. They are a profane encroachment on her ageless, dreaming tranquillity and eternal repose; her enigmatical, perhaps profound philosophy of *nitchevoo*.

Fundamentally, Archangel is a primitive center of primitive beings. Instinctively, it is a dirty hole. Hopelessly, it is a filthy place, where noxious stenches greet the nose and modern sanitation is unknown.

In the days of peace, there were perhaps three hundred fifty thousand people in the province, and sixty thousand of them dwelt in Archangel. The only other cities of importance are Pinega, with three thousand persons, some one hundred miles to the east, and Shenkurst, two hundred miles south on the Vaga River, where there were four thousand. But as a whole, the inhabitants are moujiks, dwelling in little villages of two or three hundred log huts, that in structure and design bear close resemblance to the cabins of our frontier civilization.

About these villages, the peasants have cleared the forest for a few hundred yards, and in the brief, hot months of the midnight sun, they raise meager crops of wheat and flax and potatoes. When winter comes, they are continually indoors, gathered about great ovens of fireplaces, and long through the dismal, cold, black days they sit and dream, or merely sit. They are unsophisticated folk, incredibly ignorant, but gentle, quiet mannered, sweet natured souls, despite a harsh, uncouth life; and very responsive to kind treatment.

Cholera visits them with recurrent, devastating plagues, and takes fearful toll, for they live in the midst of nauseating squalor, with total disregard to sanitation, and drink from surface wells, that in the sudden spring are reservoirs of sewage and all manner of obscene refuse.

All along the rivers and roads of the interior, at intervals of five to ten miles, are strung these moujik villages.

There is, among these people, no agriculture as we practice it in our country, with a set of prosperous looking farm buildings for the cultivation of two hundred and five hundred broad, fertile, American acres. In Russia, I never saw more than five hundred cleared acres for an entire village.

Yet, from these small, unfecund patches, the peasants, somehow, wrung the means of sustaining life, and those who toiled in the fields divided the scanty harvest with the aged and the weak, and the children who were fatherless; so that there was no mendicancy among the moujiks, and no affluence either.

There are two railways in Archangel Province, the Murman road, which begins at Murmansk on the Arctic Ocean, extends south to Kem through Petrozavodsk, and forms a juncture fifty miles east of Petrograd with the Trans-Siberian, nine hundred miles from the point of beginning; and the Archangel-Vologda railway, which reaches from Archangel four hundred miles south to Vologda, where the Siberian road comes in from Viatka on the east and leads to Petrograd. Both railways have the standard five feet gauge single track. During the winter of 1919, the Murman road, with a theoretical capacity of thirty-five hundred tons, had an actual hauling capacity of only five hundred tons a day, and its rail connections were in very poor condition and badly in need of repair. The Vologda road had a single track, but with sidings every five miles. Both roads had obsolete rolling stock, rickety, tumbled down cars and wood-burning locomotives of a type used in our country fifty years ago.

During the war with Russia, the Allies, with a medley force of friendly Russians, British, Canadians, French, a battalion of Serbians and a battalion of Italians, held the Murman railway as far south as sixty miles beyond Soroka, which is a little south of Archangel and two hundred miles to the east.

There were no Americans on this Murman railway front, except two companies of railway transportation troops, which reached Russia in April and were the last to leave in July, 1919.

Beyond the Murman and the Vologda railways, the only other highway to the interior is the Dvina, a dirty colored, broad spreading river, which from its beginning, as the Witchega, at the base of the Timan Range in Vologda province, follows a swift flowing course one thousand miles northwest to the sea at Archangel.

Sometimes, when its banks are low and it sprawls out in play, its waters glide noiselessly with a look of gentleness and peace, and the Dvina puts one in mind of our Mississippi; but usually its cold depths are freighted with grave mystery and melancholy foreboding, and then it is the spirit of Russia, hurrying by forested shores and high, desolate bluffs, where a mill, near a huddle of soiled log houses, flaps its clumsy, wooden wings, and a white church, with fantastic minaret, rears aloof, chaste and austere, in the midst of squalor.

During the period of navigable water, in the days of peace, the Dvina was plied by steamers and barges and watercraft of every description, but the freeze commences in early November, and then, until the last days of May, its waters have become a bed of thick ice.

Then, except by the Vologda railway, the only method of transportation between Archangel and the interior is by sledges, drawn over the snow by little shaggy ponies that can perform miracles of labor and seem impervious to the terrible, cold winds. These ponies are the embodiment of the moujik temperament, docile and mild mannered, very patient and long suffering, and never resentful of the most severe chastisement.

The whole province is a plain of low, gentle slopes, covered with small fir trees and several varieties of dwarfed pine. A long, dormant season and the severity of winter preclude any luxuriant, ligneous growth. Even the underbrush is sparse and thinly scattered, and commercially, about the only value of the Archangel forests is for the manufacture of pulp. The bottom of this spindly pine woods is covered with a tundra. Sometimes, there are patches of waist deep water, and in other places, a morass that seems bottomless.

Such is the character of all the North Russian forests. The natives tell stories of men, unfamiliar with the country, who have lost their way and floundered in these treacherous marshes until they passed from sight without a sign of their passage.

During the rains of fall, and when summer bursts upon winter, in June, is the season of *rasputitsa*.[18] The wagon roads then are sloughs of deep mire, and little travel is attempted. The first snow falls in November and gradually mounts, until in January it has a uniform height of three feet, except in the open places where there are great drifts much higher. No thaw comes until late February, and so moving for any distance on foot is impossible without skis or snowshoes. Cold follows the snow, gradually increasing in intensity until there are January days of forty-five and fifty degrees below zero Fahrenheit.

When the wind is high and the air filled with great, white blasts, this cold of Russia presses on the diaphragm like a ponderous weight and breathing becomes a gasping effort. In the depth of winter, the sun is banished, and during the latter part of December, only a few hours of pale, anemic glimmering separates the black Arctic night; a shadowy gloaming, like shortlived, desert twilight.

18   распу́тица "season of bad roads."

Splendid, fighting men were made weak cowards by the cumulative depression of the unbroken, Russian night and its crushing influence on the spirit; for the severest battles of the campaign were fought during the cold, black months of winter time.

Preparations for opening hostilities in the war with Russia were made in April, 1918. The Allied Supreme War Council had been alert to the presence of German troops in Finland and their fanciful menace to the Murman railway; and in the quiet harbor of Murmansk, British and French battleships had been idling purposelessly since early spring. In April, one hundred fifty Royal Marines landed from the British ships and were followed in a few weeks by four hundred more, also a landing party of French sailors. On 10th June, the United States warship, *Olympia*,[19] appeared at Murmansk, and one hundred American bluejackets disembarked. These Allied forces penetrated down the Murman railway to Klandalaksh, some two hundred fifty miles south, and, in addition to holding Murmansk, seized the port of Petchenga on the coast of Finland.

Then the scene of intervention shifted southward, and on the 1st August, General Poole, with a party of five hundred fifty French, British and a few American marines, escorted by a British cruiser, a French cruiser and a trawler fleet, attacked Archangel, which, after a bombardment, was surrendered next day by the weak Bolshevik rear guard.

The main body of the enemy had carried with them far up the river to Kotlas and down the railway to Vologda, rations, rifles, guns and ammunitions, American manufactured. Likewise, they had seized and carried off nearly all available means of transportation; and when the Allied troops examined the vast storehouses in the harbor and at Bakaritza, they found that the Bolsheviks deliberately, systematically and with great thoroughness had stripped the shelves of every conceivable thing of value. If the object of the Archangel Expedition was to safeguard the vast munitions and stores there, it had failed signally and at the outset.

Still the enemy had fled, for, by some occult form of necro-

---

19   USS *Olympia* (C-6/CA-15/CL-15/IX-40) was commissioned in 1895, saw service in the Spanish American War, and helped win a victory at the Battle of Manila Bay. In World War One, she patrolled the eastern seaboard and protected transport ships. She transported the remains of the Unknown Soldier home from France to be interred at Arlington. She was decommissioned in 1922. She still floats, as part of a museum in Philadelphia.

mancy the Bolsheviki had now become "the enemy," and it is a major premise of the military that a fleeing enemy must always be followed up. Small heed that little was known of the strength or disposition of the retiring army. They had fled. Two forces were immediately dispatched in pursuit, up the river and down the railway; and, to augment the strength of the invaders, new troops were sent from Europe.

The 339th American Infantry arrived at Archangel on 4th September, 1918. It was composed of Wisconsin and Michigan men, mostly the latter; men from our farms and from our cities, who had been drafted for war against Germany.

Like most of our civilian soldiers, they had no exuberant ecstasy for the grim business ahead, but still possessed a remarkable appreciation of the war and its deep significant issues. And they had a quiet courage that was good to see, and a quiet resolution shorn of sentimental heroics to give their lives for their country if the sacrifice was necessary. Not one of them was deeply agitated by the emotion of "Making the world safe for Democracy," which is the desiccated war cry of the academician and never could reach the heart depths of any people; but they did feel in some vague, yet definite way, that a soulless military system, which had trampled brutal, iron-clad boots through the gentle fields of Belgium, might some day carry its hateful spate to the Michigan village or green-hilled Wisconsin farm, where an old lady with spectacles sat behind the window of a white cottage, and near lilac bushes growing fragrant in the lane a wholesome faced girl waited.

These soldiers of Russia were of the same type as our men who fought in France—no better and no worse; another way of saying that they were the best soldiers in the world. They were all drawn from the Eighty-fifth Division of the National Army, and came from all the races and shades and grades and trades of our many colored American society.

Many of them had had only a few weeks of crowded military training, and were still civilians in physique and bearing. Most important of all, they were civilian in mental constitution.

With the 339th Infantry, came the 337th Field Hospital Company, the 337th Ambulance Company, and the 310th Engineers, a splendid, upstanding, competent battalion, that in the approaching ordeal upheld the best in our American traditions, showed extraordinary power of adaptite, extraordinary resourcefulness, no matter the difficulties, were ever cheerful and undaunted, and altogether splendid.

Roughly, the entire force of the Americans aggregated forty-five hundred men. It was augmented about a month later by five hundred replacements, snatched here and there from the infantry companies of the Eighty-fifth Division in France.

That September day the Americans landed at Archangel, and the fagged engines of the troop ships *Somali*, *Tydeus*, and *Nagoya* came to rest, those who looked from the decks breathed in the oppressive air a haunting presentiment of approaching evil.

28602   Convoy of the three ships which carried first american troops to Russia. Left to right: H.M.T. Somali, flagship; H.M.T. Tydeus; and H.M.T. Nagoya. Bakharitza, Russia.   Sept. 6, 1918.

Halfway from camp at Stoney Castle, England, five hundred of the little company had been stricken with the dreaded Spanish influenza. Eight days out at sea, all medical supplies were exhausted, and conditions became so congested in the ships' quarters that the sick, running high fever, were compelled to lie in the hold or on deck exposed to the chill winds.

At Archangel, there was little improvement. Soldiers were placed in old barracks, there they lay on pine boards. They had insufficient bedding, and for warmth had to keep on their clothing and boots. In this way many died and many more were enfeebled for many months, but "stuck it" with their companions and went to the front.

Had the Fates placed a curse on the Expedition from the beginning?

There was an air of inscrutable haunting sorrow in the lowering skies, glinted limpid with a sinister, bronzed light from a sun that flamed to crimson death among the dark trees over the bay.

Across the harbor projected the tiny red roofs of the city, the venerable cathedral, ghostly with great white dome, grotesque fantastic spires and minarets, garish in the fading light with startling pigments of green and gold. A mournful stillness brooded over a scene weird and alien to the men from far off Michigan and Wisconsin, who had a feeling that they had left behind forever the stage of tedious factory days and prosy farm life, and moved to another sphere, shrouded in mystery, filled with unparalleled, dread adventure.

Besides the American regiment, there was a British brigade of infantry nearly the same strength as the Americans, in the main composed of Companies of Royal Scots, most of them catalogued by the War Office as Category B2 men; unqualified for the arduous, exhausting tasks of an active field campaign, but fit enough to safeguard stores in Archangel, "light garrison duty."[20]

Many wore the bronze wound stripe, and many had two and even three of these honorary decorations. These war-tired soldiers, wearied to the point of cruel exhaustion, had given freely and without stint of their body on the Western battlefields for King and Country; but the great Empire was backed to the wall and fighting for her life in an insatiable conflict, she exacted the last draining

---

20   The 1916 Military Service Act passed by the British Parliament lists: "B2: Able to walk five miles to and from work, see and hear sufficiently for ordinary purposes."

dregs of their gasping strength. That these "crocked" Category B men performed prodigies of fortitude and miracles of endurance, and acted deeds of stirring, spiritual courage in this war of the Far North is a permanent tribute to a manhood that England breeds, and imperishable glory to British arms.

The French sent eight hundred and forty-nine men and twenty-two officers, a battalion of the 21st Colonial Infantry, two machine gun sections and two sections of seventy-five millimetre artillery.

On the railway front, there was an armored train, with one eighteen pounder, one seventy-seven millimetre and one hundred fifty-five millimetre Russian naval howitzer. Then came early in the campaign the Sixteenth Brigade Canadian Field Artillery consisting of the 67th and 68th Batteries, each with six eighteen pounders and tough gunners seasoned and scarred by four years of barrages and bombardments in France, rather keen for the adventure of North Russia while the fighting was on, and thoroughly "fed up" when there was a lull in the excitement.

These Canadians, in peace, had probably been kindly disposed farm folk, gathering the rich bronzed harvests of Saskatchewan fields.

But four years of war had wrought a transfiguration of many things and no longer did life have its exalted value of peace times. No, life was a very cheap affair, but, cheap as it was, its taking often made exhilarating sport. At the end of a battle these quiet Saskatchewan swains passed among the enemy dead like ghoulish things, stripping bodies of everything valuable, and adorning themselves with enemy boots and picturesque high fur hats, with abounding glee, like school boys on a hilarious holiday.

Yet there was nothing debased or vicious about these Canadians.[21] They were undeliberate, unpremeditated murderers, who had learned well the nice lessons of war and looked upon killing as the climax of a day's adventure, a welcomed break in the tedium of the dull military routine. Generous hearted, hardy, whole-souled murderers; I wonder how they have returned to the prosy days of peace, where courage counts for little, and men are judged not by the searching rules of war, but by the superficial standards of secure being; and living is soft and slow, an affair of rounding chores, with

---

21   Robert Graves wrote "the troops that had the worst reputation for acts of violence against prisoners were the Canadians."

few stirring moments to illumine the dull routine of most of us.

At the outset, the Canadians and a few inaccurate Russians were our only artillery. Two months after the commencement of the campaign, two Four Point Five howitzers, with British personnel, joined the Allied Forces, and there were several airplanes, considered obsolete for use in France, but good enough for the Arctic sideshow.

The air pilots were daring and courageous men, but, besides being hopelessly handicapped by defective machines, they complained that the forests of North Russia made definite discernment of the ground a very difficult thing. The facts are that they dropped several bombs on our own lines, and twice with tragic disaster. There was never any apparent reason to believe that the airplanes caused the enemy even passing uneasiness, but we were always agitated as their menacing drone approached, always grateful when they trailed off to distant skies.

The complete combat command of the Commanding General of the Allied North Russian Expedition at the outset of the campaign was then:

One regiment of American Infantry,
One brigade of British Infantry,
One battalion of French Infantry,
Two sections of French Seventy-Fives,
Two sections of French machine gunners,
One brigade (487 men) Canadian Field Artillery,
One armored train,
One 155 millimetre and
One 77 millimetre Russian howitzers.

There were a few groups of Russian Infantry with the Allied troops, but at the outset these did not number over three hundred men. In all, there were approximately nine thousand five hundred combat troops.

With this force, the Allied Commander proposed to engage in an aggressive campaign, to drive the enemy before him and follow up along the two main ways of ingress to the interior. Troops were at once dispatched down the railway to penetrate as far as the city of Vologda four hundred miles to the south, and other troops were sent by tug and barge up the Dvina River, with Kotlas, three hundred miles southeast, as their immediate objective. From Kotlas, there is a branch railway leading two hundred fifty miles further south to the Trans-Siberian at Viatka.

When their missions were accomplished, the Railway Force at Vologda would be nearly due east of the Dvina Force at Viatka, and distanced four hundred miles across the Trans-Siberian railway.

Beyond this stage, the Allied plan was somewhat hazy. It contemplated rather vagrantly a fusion with the Czecho-slovaks along the Siberian railway, after penetration south to this trunk line.

A volunteer brigade of these adventurous soldiers who had been Austro-Hungarian prisoners, but whose whole-souled sympathy was with the Allies, organized in their native Bohemia and Moravia, and joined General Broussiloff in the spring of 1917 to take part in the victory of Zborow near Lemberg.[22] Moving to the railway between Kiev and Poltava in the Ukraine, the brigade recruited more Czech prisoners in Russia until it had grown to the strength of two divisions.

After the peace of Brest-Litovsk, this army corps pushed forward to the middle Volga in the direction of Kazan and Samara intending to reach Vladivostok and sail from there to join the Allied Command in France.

The Soviet authorities promised them safe convoy over the Siberian railway, but instead, treacherously attacked at Irkutsk in Siberia on 26th May, 1918, and the Czechs then divided into two groups, one determined to fight through to Vladivostok, the other under General Gaida bent upon joining the Allied invasion from Archangel.

Although this last aim was not realized (and would have profited little if it had been) the Czechs performed a service of inestimable consequence to the Allies by acting in conjunction with the Anti-Bolshevik Siberian troops, and with the small Allied Eastern Expedition of Great Britain, Japan and the United States, in holding the Trans-Siberian open from Omsk to the coast, so preventing the transportation of many thousands of German prisoners back to Germany. When the Archangel fiasco was brought to a close they withdrew to their own country in October, 1919. And, reviewing the whole unproductive Russian effort in retrospect, the Czechs came closest towards a realization of the mythical "Eastern Front,"

22   The Czechoslovak Legion was formed originally from the French Foreign Legion of volunteers from Czechoslovakia who wanted to fight against Austria-Hungary. They fought against the Germans, Russians, Hungarian Communists, and fought along the Trans Siberian Railway to Vladivostok. They were betrayed by Kolchak and the White Russians.

for, while they could not engage in aggressive action, they did much by negative methods, denying Germany great numbers of returning soldiers that might have been welded into a considerable effective combat force for the Western theatres of war had they been free to enter their country from the Eastern frontier.

The hopelessness of a junction between the Archangel Expedition and the Czechs became certain at the beginning of the northern campaign, and General Poole was advised by the British War Intelligence that Gaida had been driven back in Samara five hundred miles from Viatka and could advance no farther before the commencement of winter.

Still the optimistic Allied Staff clung tenaciously to the belief that all the Anti-Bolshevik Russians could be joined, the Czechs, the Cossacks that General Denikin[23] had organised between the northern Caucasus and the sea of Azov, and a group of loyal officers of the Imperial Army with General Korniloff along the Don. It was within the Allied range of possibilities that all these scattered groups might join the British, French and Americans on the Siberian railway, and after the Staff was thoroughly committed to an offensive campaign, there arose the hope of cooperation from the friendly Russian forces in Siberia. On 18th September, 1918, at Ufa, there was a meeting of representatives from the Governments of Archangel, Eastern and Western Siberia, Samara and Vologda, which purported to form a Central government of all Russia, and to restore the Constituent Assembly.

On 25th October, this group moved to Omsk, created Admiral Kolchak[24] Military Dictator 18th November, and proposed to raise a strong armed force to purge Russia of Bolshevism for all time.

The Allied governments were quick to recognize this Omsk group as the de facto government of Russia.

It was hoped that the armies of Admiral Kolchak could get in communication with the Allied Forces working down from the Arctic.

This, then, was the culmination of the first stage of the cam-

23   Anton Denikin (1872-1947) was a White General and commander in chief of the forces in South Russia. He succeeded Kornilov. He would have captured Moscow but Trotsky allied with Nestor Makhno's anarchist army. He fought until 1920 and resigned in favor of Baron Peter Wrangel.
24   Admiral Alexander Kolchak (1874-1920) was a polar explorer and sailor in the Russian Navy. He fought in the Russo-Japanese War. He took power of White Forces in Siberia. He was captured and executed in 1920.

paign: There was to be a junction of the Americans, French and British from the North; Czecho-Slovaks, and the armies of Kolchak from the East; Korniloff and Denikin from the South. Tens of thousands of patriotic Russians were to join the colors of these armies, converging somewhere on the Trans-Siberian, between Perm and Vologda; from Vologda the way would be unopposed to Petrograd, and from Petrograd the Allied-Russian legions would move on and reconstruct the Eastern front, threatening Germany from the northeast!

There was nothing lacking in the imagination of the plans of the Allied High Command, whatever else might be said about them.

The Northern Expedition with great combative esprit set forth vigorously to traverse Archangel the whole length of the province by river and railway with two "Columns" which were even to penetrate well into Vologda Province.

Starting from Archangel, the Dvina river and the Vologda railway rapidly diverged east and west, so that at the first point of contact with the enemy, the two main bodies of the invader were seventy-five miles apart; and if their object, i.e., to reach the Trans-Siberian had been realized, they would have been four hundred miles apart on that railway.

There was no wire communication between these Allied Railway and River Forces, and of course liaison over the lateral terrain impassable swamp in fall, and a field of deep floundering snow in winter, was impossible.

As the invasion developed, the two columns of necessity operated as independent expeditions, with no attempt at establishing connection.

To reach their joint objective, the Siberian railway, it was necessary for the River Force to travel one hundred fifty more miles than the Railway Force. Moreover ice was expected during the first part of November, and if Kotlas was to be taken by the river, it was necessary to advance the three hundred miles in scarcely six weeks from the time of leaving Archangel.

When forced to assume the defensive in the late fall, the Dvina Column was nearly fifty miles in advance of the Railway front position, and the Vaga Column, an intervening force that was found necessary to prevent an enemy rearward movement on the river, was fifty miles in advance of the Dvina Column.

Lacking any effective communication between bodies of troops, the military incursion was expected to penetrate an unknown alien country, where there proved to be far more hostile sentiment than friendly cooperation.

There was no reconnaissance of the country; no physical inventory of the lay of the land; no reliable military maps; no knowledge of the paths through the swamp-bottomed forests; no information of the roads. Many an early attack was lost because the frontal advance failed to get support of the flanking party that became hopelessly mired in the deep marshes and never got to the fight.

The climatic conditions were a permanent obstacle to an offensive campaign. When the snow came and the weather grew intensely cold, even if we had possessed the necessary men, it would have been madness to think of an offensive in the open. Then it was possible only to dig in and hold on.

Yet despite the intense sub-zero weather there was little trouble with the field guns which during the most severe days recoiled and ran up without any jar. Moreover, there was not so much suffering from the cold as might be supposed. The Command thought that the Siberian railway would be reached before the serious winter set in, nevertheless the expedition was excellently well equipped for the Arctic weather. Soldiers were issued long fur lined coats, fur hats and had an abundance of other good warm clothing and plenty of blankets. The men from Northern Wisconsin and the Michigan peninsula did not mind greatly the severe winter days. There was some frost bite from unavoidable exposure, and much terrible privation in the defensive actions; but on the whole the Allied soldiers withstood the cold as well as the Bolsheviks.

The strength of the enemy was an unknown factor. So were his positions and his dispositions. There were no supports, no reserves. The base of the invading army in Russia was Archangel, a fortnight's journey from the far-most front and nearly three thousand miles from the main base in England; Archangel, in complete isolation during the six months of winter.

There were no reinforcements at Archangel ready to relieve the jaded soldiers so far away, who had to continue doing double duty and fighting against greatly superior numbers with no promise of relief. More important than the objective fact was the thought of being thus forsaken that froze the soldier's heart and numbed his brain and never left him through the long blackness of the days. It was the same feeling of palsied hopelessness that comes over the city bred man who finds himself lost in the wilderness. The soldier felt he was abandoned by his country, that he was forgotten and left to his fate in the grisly plain of pitiless, white Russia.

Then there was no diversion, no break in the gloomy, monotonous, despairing hours; no relaxation from the ceaseless vigilance in the guard against surprise attack; no respite from the constant threat of annihilation. The drear, sorrow freighted clouds menaced death. There was the message of Death across the bleak, endless, desolate snows. Death haunted the shrouded, hopeless days, and in the shadow of the encircling forests, Death waited. It was the most severe strain to which human intelligence could be subjected.

Many lessons were learned in the war, and none so clearly as the one that human endurance cannot be taxed beyond capacity without a resultant of diminishing military returns.

In France it soon became a corollary, universally accepted by all the Staffs, that men could not be subjected to the strain of continuing horrors and uninterrupted drain of physical resources without a pronounced lowering of fighting morale. It was calculated to a nicety how long a soldier could endure mental shocks and suffer hardships until his nervous system snapped and his distraught brain could tolerate no more.

These things were all weighed in the precise scales at the laboratories of the war establishment and provision was made for human limitations, so that there grew up three units in every combat army. One of them attacking, or standing the brunt of enemy assault; another in the supporting trenches, to be used in great emergency, but most important of all to become accustomed to the terrifying effect of the big guns; and a third that was far back, where there was a warm bath and clean clothes, peace in the sky and the soft grass still grew green, where men drank deep their little day of life, and found oblivion from the animal filth and unspeakable griefs, the awful hideousness of modern warfare. It came to be recognized that reliefs of troops on the combat first lines were as necessary as ammunition and ration supply.

But there were few and in some cases there were no relief for fighting men in North Russia, because there was no support unit from which to draw reliefs, and no reserve unit to call forth from the rear for those at the front.

The Russian Expedition, if its object was to drive the Bolsheviks clear of Archangel Province and south of the Siberian railway, required for execution of this object an army corps with entire component of artillery, and in this war with Russia, Great Britain and France and the United States failed because of:

1. Inadequate forces in the Allied Command.

This was not only true with respect to numbers, but also with respect to armament and equipment.

We had no artillery support. We were outgunned from the outset and continued to have marked artillery inferiority throughout the campaign. Time after time, the infantry, after gallant success, was shelled out of position, while our own guns were silent because outranged. The effect on the morale was most disastrous.

On the River Front, there were three Allied gunboats which cooperated effectively during the first days, but during the latter part of October, when the fight began, these withdrew to Archangel in fear of becoming caught by the ice which formed at the mouth of the Dvina, and then moved slowly upstream against the strong current.

It took a week for this ice barrier to travel one hundred miles against the course of the river, so that the enemy had unhindered opportunity to bring up his artillery mounted on watercraft, which he did, and blasted our positions for two weeks after the Allied boats had gone back to winter quarters.

Nothing was more discouraging than this hopeless inferiority in long range guns. Assaulting troops, no matter how spirited and courageous, cannot hold their advance in the teeth of a bombardment that scatters emplacements like chaff before the wind and shocks men into a state of insensibility. The stunning effect of massive, high explosives is more important than the casualties caused by direct hits. Nerves are palsied, then fly from control under unremitting blasting salvos. Fortifications are blown to atoms, and debris thrown up like vomit in a deafening belch, a bolt of hottest hell; while the earth quivers like a frightened living thing. And if modern warfare has demonstrated one thing more than any other, it is the prime, necessity of artillery support, especially during the attack. After three years' experience, the French and British Staffs laid down the rule that for an offensive to be made with any hope of success, there should be a field gun covering every ten yards of the objective and a heavy gun every thirty yards.

The British provided fifty-six heavy guns and howitzers per division, and of these twenty-nine were six inch and over.

The French had fifty-eight guns in each division, forty-six of which were six inch and over.

These divisions were made up of two brigades of two regiments each, a total of fourteen thousand four hundred men.

The Americans in France had two regiments of 75 mm. guns and one regiment of 155 mm. guns for every combat division on the first lines. At Archangel there was not a six inch gun in the Allied Command until the late days of spring when the Americans were evacuated. There was only the Russian naval howitzer on the armored train. And the only other heavy guns were two Four Point Five howitzers of the 41st Royal Field Artillery.

Besides this fatal lack of artillery, the Allied Command was miserably supplied with other armament. In the early days we had only a few machine guns and these were Vickers, with water cooled system, that became frozen and would not function in the severe cold. We had few Trench Mortars and no rifle, grenades or hand grenades. But most disheartening of all were the Russian rifles issued to the infantry. They were manufactured in our country by the million for use of the Imperial Army; long, awkward pieces, with flimsy, bolt mechanism, that frequently jammed.

These weapons had never been targeted by the Americans, and their sighting systems were calculated in Russian paces instead of yards. They had a low velocity and were thoroughly unsatisfactory. The unreliability of the rifle, prime arm of the infantry, was an important factor in the lowering of Allied morale.

2. Underestimation of the enemy forces and his military capacity.

The Allied military authorities looked with contempt upon the Bolshevik movement, and viewed it as simply a sporadic outburst of outlawry that would pass like all disorganized brigandry.

The facts were that this war was waged against the government of the Russian people. The de facto authority was in the hands of Lenin and Trotsky at Moscow. The Omsk group was distinctly an expression of the minority and the ancient Imperialists who were obstinately impervious to the new Russia flaming in revolution against age long abuses and tyrannies of the old order that could never be returned. The Omsk group never quickened any popular response. It lacked essential authority. The spectacular success of Admiral Kolchak before Perm was not followed through, and his government waned while the Bolsheviks grew in strength every day.

The Soviet army was despised as an undisciplined rabble, without equipment or officers or commissary organization. But the Bolshevik soldier was as well equipped as we were, and incomparably superior in the larger arms. He was often better rationed, and sometimes led better.

During the winter of 1919, Trotsky, an outstanding military genius, raised from the Kerensky rabble an army of one million men, which William C. Bullit[25] of our State Department saw in March of that year at Moscow, and described as thoroughly soldierly looking, thoroughly trained, well rationed, and well provided for.

From Moscow to Vologda, is less than three hundred miles by the railway which continues straight to Archangel. Why the Soviets did not concentrate a division on the railway, move straight to Archangel and leave the scattered Allied battalions bottled up in the interior is one of the many mysteries of the Expedition.

In February, Omberovitch, the Commander of the Bolshevik Northern army, announced that he would hurl the foreign invader into the White Sea and concentrated over seven thousand men in an attack on Shenkurst, the Allied position on the Vaga river. This force was ten times the strength of the defenders, who were driven back verst by verst over the deep snows to Kitsa, sixty miles down the river, and the Allied Staff prepared rearward positions in anticipation of withdrawal about Archangel and a last stand there a few weeks later. The enemy struck again with overpowering numbers at Bolshie Ozerki near the Railway.

But he never consolidated his success. For some inscrutable reason withheld the knockout blow, and, before he could reorganize for another advance, spring came with the nasta or thaw, and

25   The 'Bullitt Mission to Russia' led by William Christian Bullitt (1891-1967) was an attempt to end Allied intervention in the Russian Civil War, and to discuss other issues like war debts. His mission was a failure, primarily because of domestic political pressures on Woodrow Wilson. He publicly resigned in 1919, writing he planned "to lie on the sand and let the world go to hell." He spent time with Freud in the 1920s as his marriage fell apart. His wife had a very public affair with a painter. He served as American Ambassador to the Soviet Union in the 1930s.

he had to pull back his artillery or abandon it in the bog. He also brought great forces in November to the assault of the River position, and attacked the Railway in spring with large numbers and with great vigor; but despite his vast superiority in guns, and his great advantage in strength, he could not, or did not, break through to complete victory and destroy our scattered, weakened battalions.

Perhaps one reason the Bolsheviks did not massacre the puny Allied forces was because the nature of conditions in North Russia did not permit the concentration of great masses for the attack. The little villages, even with greatest crowding, could only house a few hundred men. Except at Shenkurst, where the most ambitious thrust was made, there was shelter for only a few thousand soldiers, and shelter was as essential as rations in this war of the Arctic.

Another reason may have been that Lenin had sagacity and imagination enough to know that a complete massacre would have fired the people of Great Britain and France and America with burning indignation and a demand for revenge which their governments could not deny. Better to whittle away the little Allied company by methods of attrition. There was no prize in Archangel. The Bolsheviks had stripped that city of everything valuable long before the Allies came to Russia.

3. Ignorance of the military commitment.

The difficulties of conducting an offensive campaign in Archangel province were at the outset not understood or realized by Allied Headquarters.

Military men have asked me why the Commanding General did not, if determined upon an aggressive warfare, concentrate his small numbers for an advance on the Vologda railway, leaving a cordon of well fortified outposts about Archangel, sufficiently distant to protect the city from artillery bombardment.

By such a method, he could have held his little force well in hand, would have safeguarded Archangel and fulfilled the real mission of the expedition (if guarding Archangel was the mission), with small cost and few casualties.

The answer to this is that British Headquarters was determined upon an offensive program, and committed itself to a punitive chase of the Bolsheviks, regardless of the nature of such an undertaking, heedless of where it led, blind to consequences.

As the Allies pushed into this unknown country, it became

apparent that between the two Columns advancing by the Dvina river and by the railway, there stretched a great, unsounded territory, entirely unreconnoitered, and through which by many routes, the enemy could threaten the tenuous unguarded lines of communication with Archangel.

It was necessary to put out flanking parties and to keep an eye to the rear. At Kodish, fifty miles east from the Railway and also on the Vaga river, which forms a junction with the Dvina one hundred and fifty miles from Archangel, it was imperative to organize invasions auxiliary to the two main bodies. Likewise, from east and west, threats were made upon the security of the city of Archangel, and it became necessary to establish detached out-posts in Pinega Valley, one hundred miles on the left flank, and Onega Valley, about the same distance on the right flank.

Also, isolated garrisons were installed in villages in the rear—at Seletskoe on the Emtsa, and at Emetskoe, where this small tributary flowed into the Dvina; at Morjagorskaya, midway between Emetskoe and Bereznik, and Bereznik itself, fifty miles farther south on the Dvina, where there was an important subsidiary base; at Shred Mekrenga, where there was an important road, and at other villages in the interior, little groups of soldiers were stationed, and often lieutenants short from civil life found themselves "Officers Commanding," faced with the problems and responsibilities of Field Officers.

By December, the Allied fighting forward stations in Archangel Province were extended in the form of a huge horseshoe, and a line drawn from flank to flank and covering the forward position would have reached out five hundred miles.

There were six principal American battlefronts: Pinega, Onega, the Vologda Railway, Kodish, the Vaga River, and the Dvina. Each of these in the war of North Russia formed a distinct episode quite apart from the others. The soldiers on the Dvina were entirely in ignorance of the fate of their companions on the Railway. At other points in the interior many did not even know that there were American outposts at Onega and Pinega; and so the history of the expedition must of necessity be a series of disjointed apparently fragmentary accounts of each separated battleground—in truth a description of six little campaigns with only one point of contact, that all Americans went out from Archangel in the fall of 1918 and in spring the following year those who still lived quit (under orders), from the same quarter.

Twice during the expedition an attempt at liaison was made between the Railway and its theoretical supporting flanks, Onega and
Kodish, and Shred Mekrenga, but both occasions demonstrated
that cooperation was impossible. The other forces on the rivers and
at Pinega were as unrelated as if they had been situate at opposite
poles. Each operated an independent, unconnected war, learning
about the other fronts only through wild and distorted rumors of
disasters, and hearing from far off Archangel only intermittently.

Thus the Allied North Russian Expedition melted away in the
snows, and the first flushed extravagant egoistic ambition of conquest and aggression was followed by a sober appraisal of the grave
peril of annihilation.

When the policy of aggression had been carried so far that it
was too late to change, General W. E. Ironside assumed command.
He was a great tower of a man, the embodiment of soldierly force
and resolution. He directly announced that all ideas of a further
offensive were abandoned and that all fronts from thenceforward
would be content to hold their ground.

General Ironside has been criticised adversely for not with-
drawing his scattered troops to Archangel to await the wreaking up
of ice in spring, when ships could enter the harbor and the fiasco
be terminated by evacuation of Russia But this criticism is unfair
and unwarranted.

It was too late for such a change of policy. It would have been
disheartening to the defenders of these distant fronts after the costly toll of the defense to have abandoned their hard fought posts.
It would have been a giving of ground that would have heartened
the enemy and thrilled him with new life; for the Bolsheviks were
never exalted by victory, they paid dearly for every inch they gained,
and our men, except when overwhelmed on the Vaga, never re-
treated from a position which they had fortified and determined to
hold.

There were no prepared defenses on the outskirts of Archangel,
and the defensive garrisons between the front lines and the city
were far separated and inadequately fortified to withstand an extensive assault. Transportation of the retreat over the deep snowed
roads would have been beset with terrible and afflicting hardship.
There were long, cruel snow spaces between the villages that lay

along the backward way and very scanty opportunities for shelter.

The task given to General Ironside, to retrieve the North Russian Expedition, was not within the range of human accomplishment. He did the best he could with the means at hand, which was to hold grimly on until those who directed from far off Europe, and who knew nothing of the gravity of the situation, or did not appear concerned if they did know, came to some sort of decision.

General Ironside conducted his defensive campaign with inspiring leadership, with unfailing heartsome courage; and he won the sympathy of all by his rare tact and understanding, and the affection of all by his consideration for the men, his efforts to stay the casualty lists.

4. The want of a definite moral purpose.

Since the days of Thermopylae, the effect of spiritual stimulus upon the fighting qualities of fighting men has been known the world over. The military people make a concrete thing of this, and attempt to diagram it, analyze and classify it in their treatises, where they call it morale.

As well might one try to reach out and touch any other manifestation of the soul. This exaltation that comes over soldiers and makes them glad to die, firm in their faith of the sacred character of their cause is above all finite measurements.

It is the purging light of the spirit that floods men's souls and raises them aloft from the restraining imprisonment of physical being to the heights of the gods. On no other grounds can one explain the superhuman valor of the lone Cheshire Company of the "Contemptibles," which, in the retreat from Mons, held up until dusk a German column of three battalions.

The French had morale at Verdun when they said, "They shall not pass," and fulfilled the eloquence of their words by the offering of their bodies.

The Americans had morale at Château-Thierry.[26]

The British at Mons, the French at Verdun, and the Americans at Chateau-Thierry, fought as they did because they knew, or thought they knew, the cause of the fight.

But in Russia, the soldier was never told why he fought. At

---

26   The Battle of Château-Thierry was fought in May and July 1918, the first real action the AEF saw. The Germans had pushed the French back 40 miles, and after blocking that advance the Americans launched a counter-offensive.

first, this was not thought necessary. Then the High Command, remembering the importance of morale, and recognising the need of some sort of explanation, if only for the purpose of regularity when men were asked to risk their lives, issued proclamations that puzzled and confused the soldier more than if a course of silence had been followed.

While all this time to the Americans came newspapers from home with accounts of speeches by politicians and demagogues who fired Bolshevik bullets from the rear and extolled the Soviet cause, hailing it as an heroic progression in human effort.[27]

There is another axiom in the military books, that soldiers fight best on their native soil and in defense of their homes; but here was a company taken fresh from civil occupations, with a civilian mental outlook, set adrift in an alien country, six thousand miles from home, engaged in a desperate, sanguinary war, and asked to undergo privation and hardship, to face untold perils for unmentionable reasons.

Still, though the expedition was committed to no definite moral purpose, there was a morale in North Russia. A morale that arose from comradeship in a fated enterprise, a morale of seeing the bitter game through, taking risks and meeting perils that must be borne by others if even one shirked his share. A noble, selfless devotion, playing the man's part in a lottery with Death, where Life was the stake. The upholding of some elemental metaphysical creed that could be definitely felt but never understood, a code of challenged manhood that had come down through many generations of warring ancestors—this was the morale of North Russia; it brought forth the best and the purest in our manhood, and recorded deeds that no survivor can recall without quickening heart beats, and a profound belief from what he saw, that the spirit is supreme and triumphs over the body of man.

5. The Russian people did not rally to the Allied Cause.

If the fight was for Russia, the Russian people were cold and apathetic, the worst of ingrates. Many Russians had the impression that we had come to restore another Romanoff to the throne.

The statement of the American government, with respect to the reasons for military intervention, put the case as if the Allies were

---

27   Out of ignorance, political sympathy, or countersignaling spite, many Americans supported the Russian Revolution. Indeed, the 'American goes to Soviet Russia and sings its praises' is a subgenre of the time.

engaged in a highminded, selfless service for Russia, but the great mass of moujiks were indifferent to our immolation, and showed undisguised relief when we finally and ignominiously quitted their Country.

During early August, a government of the north had been installed at Archangel by a coalition of Cadets, Minimalist, members of the People's Party and Social Democrats, with a bourgeois cabinet and with an old man, Nicolai Tschaikovsky,[28] as President of the province. But it was a fact known to all that the Allies determined the policies of this government, that it was in fact merely a guise for an Allied Protectorate.

This government of the North it was that had invited military intervention; but had a plebiscite been called, the people would have registered their voice in unmistakable terms and volubly Russian "Let us alone. *Nitchevoo.*"

Thus the campaign was another effort of England to impose her will upon an inferior people, and bring them for their own good to a higher order of things, disregardful of their volition in the premises. It was an echo of South Africa and Egypt, Mesopotamia and India, inspired by that lofty faith in Britain and the immortal commission of the Empire to rule an afflicted world and bring the blessings of sustained order, where only trouble and chaos prevailed before.

In Archangel, an ambitious attempt was made to recruit Russians under the high sounding name of The Slavo British Allied Legion, and after most energetic efforts, about two thousand starved moujiks, seeking something to eat, joined the ranks; indifferent mercenaries never to be trusted in the tight positions. They were given the khaki of the Tommy, but there all resemblance to the British men of war ended. Their pay was in worthless rubles. They were given an inferior ration, were treated patronizingly. Between them and the Allied soldiers there never was that generous comradeship that leaps the restraints of divergent language and manners when men fight shoulder to shoulder for life and some things that are more dear than life itself. It was a case of alien spirit above all else. British officers never could understand why the Russian officers, with the acute, sensitive nature of the Slav, were quick to feel and keen to resent, seemingly studied slights and snubs

---

28   Nikolai Tchaikovsky (1850-1926) was an old school revolutionary. He worked for the Russian Red Cross. He tried to unite leftist factions against the Bolsheviks. He died in exile, in England.

and discourtesies. Russians of culture and refinement never could penetrate the unfailing reticence and frigid unsympathetic exterior in which gentlemen of England have been schooled for generations beyond memory, habitually to conceal the emotions.

When the utter failure of the volunteer system became certain, thousands of Russians were coerced into the army by a draft system; but these failed too, because their hearts were cold to Russian patriotic British appeals; because there was no great moral issue, no moving cause for the fight.

The war with Russia was in fact a typical British show, conducted by that conquering people who have spread the dominions of the mother country to every shore of the far seas. A war that was waged with the invincible will, that noble effacement of physical comfort; that indomitable purpose and masterful determination; that courage and careless naivete, and contempt of danger and risk; that splendid sportsmanship, that love of fair play; and all the sublime self sufficiency, all the muddling, blundering and fuddling, the lack of understanding, the brutal arrogance and cold conceit, and apparent heartlessness and want of sympathy that are forever British.

Naturally, the British assumed direction, just as in France when the first Americans came Clemenceau and the Earl Haig demanded that they be fed piece meal to the French and British front divisions; but the soldier, Pershing, sensing the important moral value of having his men go to battle under the American flag and directed by American officers, waited and would not yield to the strongest pressure. And it was an American army that brought us to glory at Saint Mihiel and Chateau-Thierry and the Argonne forest; an all-American army led by American divisional commanders.

There are racial differences, racial prejudices, racial disparities, and racial asperities that cannot be gainsaid even under the influence of impersonal military discipline, and experience has shown that soldiers yield a more ready obedience to leaders who speak their own language; understand the philosophy of their daily lives, and at no other time did General Pershing so demonstrate his greatness, his complete understanding of the perplexities in Allied military organization as by his courageous insistence upon the solidarity of the American army on the battlefields of France.

But in Russia the American regiment was at once merged with the British Command, and from first action until the end of the campaign, British Headquarters directed and controlled the dispositions and conduct of the Americans.

At Archangel there is a modern, spacious white building, and here from steam-heated headquarters Colonel George W. Stewart[29] commanded the United States 339th Infantry, here were quartered his staff officers, the unemployed "brains" of our Northern American army. He never saw any part of his regiment in action. For a long time I believe he had not even a vague notion regarding the location of his British dissipated troops.

Embassies of France and Serbia, Poland, and Italy were in Archangel, and the American Ambassador, David R. Francis, came from Vologda there early in August, and stayed until sickness compelled him to leave for England during the winter. And there was an American Military Attache who developed into a Military Mission with Colonel James A. Ruggles as chief, and a staff of officers to assist him. Also there was an American Consulate, with an American Consul General, Dewitt C. Poole, who at times appeared to take over a supervision of the American share in this strange, strange war with Russia.

And over across the harbor at Bakaritza, a well-fed Supply Company watched over mountains of rations and supplies that had been brought all the way from far off America; supplies and little good things and comforts that would have heartened and brought new life and hope to the lonely, abandoned men on the far fighting lines in the snow. These supplies never reached the front, but the Supply Company, with American business shrewdness and American aptitude for trading, acquired great bundles of rubles, and at the market place converted these into stable sterling, and came out of Russia in the springtime with pleasant memories of a tourist winter; likewise a small fortune securely hid in their olive drab breeches. But there were others who ate their hearts away, fretting and chafing, in Archangel, whose petitions to go to the front to play the man's game were denied by those in command.

British G.H.Q. brought six hundred surplus officers and forty thousand cases of good Scotch whiskey. Some of the officers had come frankly in search of a "cushy job" in a zone they thought safely removed from poison gases and bombardments and all the hideous muck of the trenches. Others, much to their disgust, had

29   Colonel George Stewart (1872-1946) served in the Phillipine-American War, and received a Medal of Honor when he saved one of his enlisted men from drowning in a river while under fire, on Pinay Island.

been sent to the polar regions because someone in Headquarters had thought they possessed some peculiar qualification to command or "get on" with imaginary Russian regiments that were to spring to the Allied Standard.

So it was that Archangel became a city of many colors, as gallant, uniformed gentlemen strode down the Troitsky Prospect, whipping the air with their walking sticks, and looking very stern and commanding, as they answered many salutes, in a bored, absent-minded way.

There were officers of the Imperial Army, weighed down with glittering, ponderous honor medals, and dark Cossacks with high gray hats, and gaudy tunics, and murderous noisy sabers. Handsome gentlemen of war from England, from Serbia, Italy, Finland, France, and Bohemia, and many other countries, all arrayed in brilliant plumage, and shining boots, and bright spurs, and every other kind of "eyewash." And, of course, there were large numbers of batmen to shine the boots and burnish the spurs, and keep all in fine order, and other batmen to look after the appointments of the officers' club, and serve the whiskey and soda.

In the afternoons there were teas, and receptions and matinees, and dances in the evening, when the band played and every one was flushed with pleasure and excitement. Such flirtations with the pretty barishnas,[30] such whispered gossip and intrigue and scandal in light-hearted Archangel!

At Kodish, at Onega on the Vaga, and at Toulgas, far off across the haunting snows, sick men and broken men, men faint from lack of nutrition, and men sickened in soul, were doing sentry through the numbing, cold nights, because there were none to take their places in the blockhouses, and no supports to come to their relief, no reserves to hearten them and give them courage.

The blockhouses so far away, where men were maimed and crippled and shell shocked, and the black hopelessness that crept into men's hearts, and strangled men's hearts, and overcame their soldier spirit—in the blockhouses — far, so far off from gala Archangel.

---

30   Барышня, young lady, lass, miss, i.e. a young woman. Sovietskaya barishna, young soviet ladies. Perhaps the author suggests the young ladies were a great comfort to the soldiers.

28599   One hundred and twenty three Bolshevik prisoners, captured by the
         French fighting along the railroad between Archangel and Vologda
         being counted on their arrival at Bakharitza by an American sold:
         and a sailor from the U.S.S.Olympia. The prisoners were brought
         to Archangel under guard of American sailors.   Bakharitza, Russia,
         Sept.6,1918.

THE RAILWAY

"We are not declaring war, nor making war on the Lenin and Trotsky government, because it is not our affair."

SENATOR HITCHCOCK, Chairman of Foreign Relations Committee in the Senate of the United States. 13th February, 1919.

THE RAILWAY

WHEN the troops of Poole's first expedition divided at Archangel, and one group was sent up the Dvina; another which was a part of the French Colonial battalion was told off for pursuit of the Bolsheviks down the Archangel-Vologda railway.

Hot and eager for first blood, the French hurried forward until the Kayama River was reached, where the enemy made an unexpected stand. There was a sharp engagement, the Bolsheviks were severely punished, and one hundred and fifty prisoners fell to the Allies.

But a little further, at Obozerskaya, some hundred miles south of Archangel, the despised fugitives turned again and displayed an amazing disposition for combat, entirely at variance with the cowed spirit of the feeble rear guard that had surrendered Archangel.

They came back in force and greatly outnumbered the Allies, and there was in the defiant attitude of the Red troops reason to believe that the Soviet chieftains had taken stock of the military situation, had verified the preposterous intelligence that the Three Great Powers—Great Britain, France, and the United States—were definitely bent upon war and seriously intended to invade the great domain of Russia with scarcely two infantry combat regiments!

Reports came of fast gathering Bolshevik armies at all fronts massing for attack, prepared to take offensive action on a grand scale, and, hardly had the campaign entered upon its initial phase, when the utter inadequacy of General Poole's numbers made egregiously evident the impossibility of the proposed investment by River and Railway.

The two "Columns" were in simple truth little patrol parties, and, as they drove further into the interior, the ridiculous audaciousness of their ambition to sweep the enemy from Archangel Province, and south even beyond Vologda Province, seemed almost beyond the purview of sane contemplation.

Highways for flank envelopment, and byways for encirclement, commenced to appear with discouraging frequency the further the advance developed in this unknown, speculative, shadowy hinterland, and all of these avenues for surprise attack had to be watched and safeguarded. One of these was the Vaga river, which meets the Dvina near the Allied subsidiary base at Bereznik; where an auxiliary, flanking expedition was detailed from the River Column, for this tributary is capable of floating substantial craft that could transport artillery and many infantry from the Bolshevik stronghold at Velsk in Vologda Province, and north of Velsk is Shenkurst, the second city of Archangel, with a political significance that could not be neglected by this politico-military excursion into the interior of Russia.

If left unguarded, the Vaga would be an open invitation for the Bolsheviks to capture this supply depot, Bereznik, and gain the rear of the Allied Dvina forces.

Many other routes for enemy movement developed as the invasion paused, undecided whether to retire for consolidation, or to try to plug up these many openings for enemy movement, and as the Command stood hesitant, still other approaches by flank and rear were revealed.

It was (or became) known that the headquarters of the Sixth Bolshevik Army was stationed at the city of Vologda, from which its commander could send troops north along the railway, and assail the Allied frontal position, or detrain, and move his men on roads and trails that took off along this route and led to the Allies' flanks and rear.

One of these roads follows down the Onega valley north to the port of Onega.

At Chekuevo, it is nearly opposite the Allied advanced railway position, Obozerskaya, and these two villages are joined, fifty miles cross-country, by a good roadway that in winter is capable of supporting artillery carriage. Some fifteen miles west from Obozerskaya, on the same road, Bolshie Ozerki, several groups of moujik huts, lies in sprawling confusion.

Late in the winter, a pitiful little outpost of French and friendly Russians, an immolation to this campaign of invincible folly, was destroyed at Bolshie Ozerki in a massed enemy effort that sought to annihilate the whole Expedition.

A few platoons of American infantry were stationed at Onega to shield Archangel from the west, and to watch this Onega, Chekuevo, Bolshie Ozerki, Obozerskaya communication line, which linked up Archangel with Murmansk, and, during the frozen months, was the only outlet to the world beyond the Arctic Sea.

The main Bolshevik stronghold north of Vologda was at Plesetskaya, some fifty miles south of the furthermost position of the Allies on the railway, from which an Imperial Government highway reached out through Archangel Province northeast as far as Emetskoe, on the Dvina, passing through the villages Kochmas, Avda, Kodish, and Seletskoe, near the Emtsa river. At Kochmas, another road branched east to Tarasovo, thence north through Gora and Shred Mekrenga.

From Shred Mekrenga and Seletskoe, the enemy could have access to the lower Dvina, head off all supply convoys for the Dvina and Vaga columns; and hold the Allies trapped far up stream. Therefore, two more auxiliary expeditions were organized, and, instead of two invading "Columns," the Allied forces, woefully insufficient at the outset, were operating in seven columns, separated detachments, advance parties, outguards, outposts, flanking forces, and all along the Dvina, from Kholmogora to Bereznik, a stretch of one hundred miles, were still other detached soldier groups watching the treacherous way from Archangel, a Cossack Post in one village, a squad in another, in still another a platoon, all without communication and completely undefended in case of real attack.

There was unlimited chance for rear movements along that tenuous, unprotected, communication line. General Ironside would have massacred the Bolsheviks had positions been reversed. The Germans would have annihilated the Allied North Russian Expedition with half the numbers that the Bolsheviks had.

During the winter, several circling movements were essayed, but never on a scale of comprehensive organization; at Morjagorskaya, in February, and at Shred Mekrenga, the enemy came closest to success, but at both places was stopped by the gallant British, and when spring came his chances vanished, the bogging quagmire precluded any further offensive. But while the Bolsheviks did not destroy the Expedition, they soon reduced the invasion to a series of desperate, detached, outguard actions, and the River and Railway Columns that were to have entered Kotlas and Vologda with the coming of the first snow, were flung far and broad over vast Archangel, as the effort "to stage a real show with two men and an orange" wilted with the first snow, a dismal, ghastly "washout."

Even when the Americans reached Archangel in September, the campaign had already assumed a defensive character. Indeed, so serious was the outlook that they were rushed from the troop-ships, shunted off to Russian box cars, and consigned with expeditious haste to the Railway Front.

Nothing of this was known to these new zealous soldiers off from a brief military training encampment to the very heart of war's purple, glamourous adventure. And it is doubtful whether they could have realized the significance of the military situation, even had it been communicated to them. In a few crowded weeks, so many stirring events had thronged their heretofore placid lives that these recruits from Michigan and Wisconsin were buried beneath a bewildering wilderness of amazing impressions through which confused, alien scenes and persons and places trooped in phantom and fantastic multicolored parade, until their minds were stunned beyond the power of further reception.

During the long voyage, a few still civilian in mind, had recovered sufficient equipoise to inquire about the connection between a war in Russia against Germany, but the inquiry was so unproductive, so futile, and there were so many eccentric twists and turns to this stupendous world madness that in most part they soon fell into that fatalistic philosophy of all soldiers; most of them were content to place their unbounded trust in those who sat in the high places and whose omniscience guided from afar. It was far more quieting, vastly more satisfactory.

Once, during that swaying night journey, from Archangel to the battle line, the decrepit Russian locomotive gasped convulsively and stood still by an old station of huge logs, and, under the lurid light of a flaming torch, was revealed a trainload of prisoners, passing north from the scene of hostilities somewhere below. They made an unheroic spectacle, with their shrinking countenances and unsoldierly, nondescript uniforms, so that some American wag, in a spirit of bantering patronage, called them "Bolo wild men," a name that clung to the enemy throughout all the days of the campaign.

But the shabby prisoners, first living sign of real battle, sent a thrill up and down the spines of these young men, who were so ardent for war and knew so little about it. They sniffed the air of conflict, yearned to give the "Bolos" a taste of their quality, and promised themselves that the folks back home would have nothing to be ashamed of when they came under fire.

The next morning the depressing aspect of the dirty, unkempt group of huts where the soldiers detrained almost passed unnoticed alongside the captivating spectacle that stood on the track nearby, a ferocious war monster, with massive plates of steel like dragon's scales, huge funneled naval guns, and locomotive set in rear of trucks which were piled with sand bag barricades where Lewis automatics poked out murderously, manned by a hodge-podge Polish-Russian crew, who were themselves manned by competent appearing, war-weathered British N.C.O's.

A narrow threadlike swath trailed through the stunted starveling forest to the lowering gloom of dull, laden skies, and the hearts of the fresh, battle eager soldiers swelled big as they gazed far down the gleaming rails to the murky mystery of No Man's Land.

There was in the air a peculiar, dispiriting quality, a brooding, pensive, Russian note that cannot be made known except to those who have felt it. Stillness, heavy almost to the point of suffocation, the shroud of skies that hover mourning on the trees, and the shadow of unlifted gloom that reaches out from the forest and bears down upon the spirit with deep intangible melancholy.

Suddenly the quiet was broken by the distant boom of a heavy gun. Then an ominous whine circled from the ground, approached snarling stridently high in air, and fell with a crumbling roar seemingly very near the new soldiers, who, on command, scampered to cover from their erect column of twos on the naked embankment.

A cordon of strongpoints had been constructed around the village, Obozerskaya, and these the Americans took over, tensed for the impending battle.

But inexplicable days passed, and the Bolo did not come. There was not even a feint of attack, and the Allied Command, with short memory for the hazardous nature of its extended position, the apprehension it had felt only a little while before, began to chafe for action, became impatient; again the military fetish of an "offensive campaign" grew, waxed strong, became assertive once more, and again the ambitious vision arose to take Vologda before the snow.

"All patrols must be aggressive," directed a secret order of the officer in command, "and it must be impressed on all ranks that we are fighting an offensive war and not a defensive one."

So American officers, directed by ranking British officers, moved their companies forward to the "offensive war," and four miles beyond Obozerskaya, where a post on the railway bore the Russian characters "Verst Four Sixty Six," they closed with the Bolos and drove them beyond the bridge at Verst Four Sixty Four.

In the counter-attack that soon followed, one platoon of the Americans, separated in the swamps of the woods, was nearly enveloped. It fought until all ammunition was exhausted, and then the officer, Lieutenant Gordon Reese, had no thought of submission. After the last cartridge was gone, the bayonets still remained, and after the bayonet, came doubled fists. At word of command, the platoon fixed bayonets, went forward with a yelling charge, broke down the Bolsheviks by their sheer courage and impetuosity, and the endangered men were able to join the main body of their comrades, repulsing the attack.

Before Verst Four Fifty Eight, Allied aggressive operations were resumed when one of the French companies came back from Archangel to assist in moving against the strong enemy works. There was a bridge at Verst Four Fifty Eight. If this was destroyed, it would take a long time to rebuild and seriously impede the "offensive war" down the Railway. It was, therefore, intended to drive the Bolos back so violently that they would have no chance to detonate the important bridge.

The plan of attack was for a three-fold movement: front, right flank and rear. The French company, supported by the artillery of the armored train, an American machine gun section, and twenty-one Americans, with three Stokes mortars (who were not entirely sure of the use of these weapons) were to hit out at front. The rest of the Americans, two infantry companies, were to form as many detachments and rush the enemy from his east flank and rear at his furthermost trench back at Verst Four Fifty Five.

The frontal assault would wait on these circling movements; a bivouac in the woods, and at dawn, timed together, the three parties would move to the three-quartered battle. The distance through the woods to the enemy rear was "estimated at from six to eight miles."

But, in execution, the plan failed dismally, like many an operation that carries through flawlessly around the military council table, for "estimates" are of little use in the service of battle conduct, where time is reckoned in seconds, and victory measured in minutely fluctuating scales.

The contemplated operation was to approach the enemy flank through one of those lofty, forest aisles, which were cut with masterful, precise woodcraft by the engineers of Peter the Great, entirely transverse Archangel Province. Regularly, narrow lanes intersect these forest aisles, and it seemed to the officer planning this attack a simple thing to follow one of these lanes, and take the course of a north and south aisle until a point was reached opposite the enemy position. He did not know that those forest paths were deep with clinging, slimy morass, and bog that gave no footing, that frequently the main cuttings opened before shallow lakes of open water. There was no reliable map to show these things, and no native would admit that he knew the way.

So the attackers went forth over unknown ground, and soon were stumbling in a blackness so dense that one file could not see even the outline of the preceding file. The sinking bog made the march distressingly arduous, yet for hours the company kept resolutely on, when, without warning, the forest parted and the sodden way terminated in a wide sheet of open water.

It is impossible in the night blindness to know position or location, or how far the exhausting, laborious pace has made. Startlingly near comes the coughing exhaust of a locomotive, doubtless the armored train standing by the Bolshevik defenses on the tracks.

In their jaded and spent condition, the men are ill fit to engage in battle, yet there is nothing to do but have a go at it, so plowing through waist deep swamp and awful, oozing quagmire, they lurch on. Struggling forward, still forward, they are caught and tripped, and sprawl splashing in the cold water and the bog, but they get up and drag on until all are breathing with heavy, sobbing gasps; and under the strain of terrible exertion, all are weakened, some so done in, that they lie in the water like wounded animals on their haunches, and have to be helped forward by others of more physical strength or greater will.

In this agonizing way, perhaps a few hundred wallowing yards are made, but it is clear that the company cannot go on, and there is no hope of end to the miserable, sinking marsh; so the officers hold council, and decide, not without great reluctance, to abandon their mission, and the word is passed on to the scattered troops to follow back over the way they came.

In the darkness and the trackless morass, this is not easy, as through the endless black night the lost company struggles flounderingly and with little hope, until the heart of all is cold with despair; but more blighting than the knowledge of being lost in the wilderness of Russian swamps, and the depression of abject, physical exhaustion, is the mordant disappointment of failing the expectant French in the coming fight.

At dawn, two soldiers, who, in days of peace, had been timber cruisers in the pine woods of the Michigan Peninsula, led their comrades to ground firm enough for footing, and half dead from fatigue, brought them back to the railway, but too late, for hours before the tumult and shots of battle had reverberated from far advanced ground on the railway tracks; for, at the appointed hour, hoping that the cooperating actions would still develop, the French went in to the attack, supported by the American trench mortars and machine guns, and smashed the enemy from his foremost lines. Directly he rallied and returned in force to the counter-attack in which many French were killed, the trench mortar section was decimated and lost most of the guns, the machine gunners put out of action, and the whole little force was shoved back over much of the freshly won ground to the bridge at Verst Four Fifty Eight, where the Americans stood with braced backs and would not yield.

For two days, the Bolo armored train showered them with shrapnel, and upcasted tons of high explosives that tore glaring, wide wounds in the railway track, till theoretically they were hammered into submission, but when the Bolshevik infantry, in the gray hours of dawn and dusk, approached to take the crucial position, they were always driven to cover by a heroic defense that never failed. So the bridge was held under difficulties that would have shaken ordinary troops and caused them to fall back, but not in Russia, for that was the way of this queer little war. Priceless lives would be lost, much blood run, and stirring exploits of courage and noble sacrifices be performed, to safeguard a little bridge like Verst Four Fifty Eight, or a dirty village that objectively meant nothing. Yet what sacrilege to have breathed this to the soldiers who bled for them; for to those who risked their lives and yielded up their lives, rather than desert some little bridge or moujik village, these signified the shibboleth of North Russia.

For inordinate stress was placed upon these inconsequential, hard contended spots; they became graphic in the imagination, cardinal precepts in some strange soldier creed, altars upon which friends had given all as proof of a comradeship triumphant over self and self desire. Indeed, with the fresh recollection of courageous comrades now dead, their abiding faith in him, and the thought of those far back at home, whose eyes watched from afar with undimmed loyalty, did he not stamp himself as a craven if he failed, a mongrel thing unfaithful to his breeding?

Thus has it always been. The race has carried on by dimly understood, irrational traditions that move men to the profoundest depths and challenge elemental impulses that have descended in transmuted ancestral determinism, we know not how or why. And if we are to endure, it must be by these same primal emotions, that cause men the world over to scorn soft ease and security for the sake of a vague, inexplicable ideal; inchoate conceptions of service; passionate, stirring impulses lacking definition, which, are born with life itself, reach down to the bottommost depth of nature and transcend all feeble efforts of analysis and artificial ratiocination.

So it came that the momentous bridge at Verst Four Fifty Eight stood fast, and the Bolshevik attack beat against an unyielding rock until it spent itself by its own fury. Then the position was consolidated, Allied headquarters moved nearly three miles down the railbed, and the dead, in order that there might be no interruption of the renewed offensive, were laid away in white Obozerskaya churchyard, beneath rough crosses of wood, such harsh emblems of life's surcease, and so fitting in this inflexible, cold, repellent north world.

After a fortnight of more scheming and preparation, the forest was carefully reconnoitered, a path that could be traversed was found through the swamps, in a three cornered attack, the Allied position advanced to Verst Four Fifty Five; and pressing on, the Americans and French went forward to still further battle. But now occurred an event more baneful to the Expedition than all the enemy attacks. The month was only October, but in some mysterious way, the French had already received word of the pending Armistice, and entirely unmoved by the disaster that might befall their abandoned comrades, the whole French company quit the front and went back to Obozerskaya in an ugly mood.

"The war is over in France," they argued, "why should we be fighting here in Russia when France has declared no war on Russia or the Bolsheviki?"

Ninety of the mutineers were placed under arrest, and returned to Archangel for confinement.

It is not known whether or not the Bolsheviks were directly apprised of the mutiny, but hardly had the French retired, when the enemy artillery laid down a shaking barrage, and when night came, the lone group of Americans were standing off a great horde of Bolo infantry that only waited for dawn to continue an overwhelming assault.

Clearings occurred at intervals of several miles all along the Vologda railway. Usually they were in the shape of large squares, a half mile or more across, with log stations, several woodchoppers' houses in the center, and near them piles of corded pine to feed the wood burning locomotives. The next day when the supports came up they nearly blundered on a large Bolshevik force massed for a surprise attack in one of these clearings.

With unerring, quick-witted appraisal, the American officer saw that he was outnumbered three to one, but losing no time, he divided his company into three parts and struck out from three directions of the woods, firing rapid fire, making a great commotion and noise, to give the impression of great numbers.

Most of the enemy troops were poorly disciplined and poorly led in these days of the Fall campaign, and this ruse of the three-cornered attack was carried through with such colored theatrical effect that it scored complete success. There was a brief fight, some good Americans shooting at open, closely grouped targets, and the frightened Bolsheviks fled in disorder. Not only were the Americans able to relieve their threatened comrades, but the scattered Bolsheviks were followed up to Verst Four Forty Five.

This was the furthermost point of the advance, for soon General Ironside assumed the office of Commander-in-Chief, and the "offensive war" was heard of no more. The campaign became a stalemate, each side awaiting the opponent's next move, and not till November did the Bolsheviks become aggressive again. Then they stormed the positions with great determination, but all posts held and they were thrown back with frightful loss.

The succeeding month, it was decided by the Allied Command to capture Plesetskaya, so that the enemy might be denied a base for winter movement, and the divergent Allied forces of the Railway brought together. But the effort failed. The Russian contingent that was to go on skis around the left, fifteen miles to Emtsa, floundered helplessly, became exhausted and funked out in the deep snow many miles from their objective; also the auxiliary force at Shred Mekrenga could not gain its ground; but most of all, the failure was caused by the members of the Slavo-British Allied Legion, who faithlessly deserted in large numbers and went over to their countrymen, the Bolsheviks, with full information of the Allied plans.

This marked the collapse of the invasion of Archangel, and when the cold of winter had settled, the Red leaders set busily about the task of planning the destruction of the over-extended Allied, lines on six unsupported fronts, which could neither retire beyond Archangel, nor be reinforced until the remote coming of spring. It looked as if the great military machine which Trotsky assembled, would speedily crush Ironside's men, and the Moscow newspapers announced that a million Red bayonets would hurl the foreigners to the White Sea, and into it (although the sea was then solid ice), but inexplicably strange, after the failure of Plesetskaya, there were few stirring, winter days on the Railway Front, except once, when a daring Bolshevik raiding excursion on skis snatched one of the rear guns from the French (who had been shamed into returning to the front), destroyed it, and got away in the snow.

Major J.B. Nichols was at this Railway Front, a civilian officer, and the only one of the Americans in senior authority who appeared to possess a heart, and courage, and unfogged discernment. He early grasped the vain futility of the whole campaign and no cajoling or flattery or threats from Archangel could sway his refusal to engage a single man in unavailing patrols through the ambushed forests or in hazardous "blow-offs" between the contested lines, that accomplished nothing save the sacrifice of life. So for the most part the winter defense was a routine of work on the defenses, the dugouts and the fortifications, and necessary reconnaissance parties over the trails, to watch the flank approaches and to keep an eye on dangerous Bolshie Ozerki.

With ready methods of quick transportation, and an increase in the garrison by the coming of the King's Own Liverpools, it became possible to arrange spells of relief, and in March the Americans went back to Archangel.

At the front it was different. There was a tautness, a hushed, dread expectancy in the air, and life, an uncertain thing, was to be lived, like the Hedonist, for the day; there was no time to analyze the causes of one's misery or even to be more than dully conscious of it; pressing urgencies, actual or imaginary, were always occurring, and they crowded out all opportunities for contemplation and introspection.

But there was no pressure in careless Archangel, where harrowing care and disgusting, swinelike filth vanished with a wave of fairy wand and lo, the war with Russia became a magical heroic pageant. Large numbers of unemployed officers strolled the Troitsky Prospect, very merry and bright, an array of bright, varicolored ribbons, like flower gardens, flourishing on their well-arched military chests.

There was the American Supply Company at full strength, which looked very sleek and smug, and groomed well, and well fortified to withstand the rigors of the Arctic winter, who displayed extraordinary capacity for trading with the natives and astounding dexterity in the acquisition of an affluent wealth of Russian rubles.

It made a soldier sick at heart to see the good things stacked high at Bakaritza, the sweets and dainties and tobacco that would have meant so much to the homesick Vaga men and the far Dvina men who were never relieved—the cases and cases of whisky piled in mountainous piles in the warehouses at Bakaritza!

There were other cases (empty ones) outside the Officers' Club. And in the happy city, parties were held, with sparkling jollity, and entertainments, and dances, and jingling sleigh rides, and down the long toboggan run near the domed cathedral roistering funmakers with screaming laughter would glide through the exhilarating Arctic air to the white world below. The varied military were having a rather unique and amusing time of it in jaunty galliard Archangel, and none of the impassive Slavs there seemed agitated or even interested in this war to bring peace to "sad, distressed, and afflicted Russia," which had ended life for many Americans and broken the lives of many more.

Russian soldiery was everywhere, Russian officers, with gaudy uniform and restored Imperialistic hauteur; and Russian soldiers drilling cm the parade grounds, with a snap and a smartness that was oddly British, all fit and well-fed looking, capable of destroying untold American rations, with the appearance of being able to shoulder a musket in defense of their country if they were so minded, but with no apparent intention of being so disposed.

Every soldier knew of the scene at Alexandra Nevsky Barracks, where American machine guns were turned on the S.B.A.Ls. to put down the revolt that occurred when our Russian allies were ordered to the fighting front. And poignantly fresh was the memory of the faithless conduct that had lost Plesetskaya in December. Treachery at the front, and treachery stabbing in rear! Why should American soldiers die and suffer exposure and hardship for these heedless, indifferent people?

And if the fight was not for Russia, what was it for?

There were persistent rumors of a war to collect imperialistic claims and money obligations, and other passing rumors as errant and disordered as the Red Bolo Bolshevik propaganda that begot them. But was it altogether strange, that after this had gone on for months and months, when the soldier asked for the facts and the facts were denied him, that he should begin to wonder, and to grow almost embittered; that, in fact, one of the companies should give audible expression to its turbulence?

During the last part of March, a convoy of sleighs drew up before Smolny Barracks to carry this company and its equipment over the frozen bay of Archangel to the station where a train was waiting to take them to the Railway Front. But the men did not stir from their barracks, and the equipment was not loaded, so that the colonel of the American regiment came (somewhat hastily) from his warm quarters to learn the reason for the delay.

The colonel assembled his soldiers in a large Y.M.C.A. hall, and read them that Article of War which pronounces death as the penalty for mutiny. Then, following an impressive stillness, he asked if there were any questions. There were no inquiries concerning the Article of War, which is terse, succinct and unequivocal, but one soldier arose very respectfully and said:

"Sir, what are we here for, and what are the intentions of the U. S. Government?"

The colonel very frankly replied he could not give a definite answer to the question, but added, that regardless of the purposes of the Expedition, it was now in acute jeopardy of extinction, and the lives of all depended upon successful resistance. More silence followed.

There is a favorite disciplinary method of the military based upon basic, elementary psychics. It is invoked by all, from the drill sergeant to the general officer. The principle is the antithesis of mob psychology, and goes upon the presumption that man is a gregarious being.

At the first rumor of incipient disorder, soldiers are assembled at attention, and any man holding to minority views is commanded to step forward (usually three paces) from the ranks and expound his convictions.

Great heroes and those capable of the highest, unparalleled

courage, quail at this test, for it is one thing to rebel in company, or in the secret counsels of one's inner conscience; quite another to stand out stark alone and unsupported against the strong arm of the military, the harsh, punitive, martial law of an intolerant warring nation, that can brook no infringement of combat discipline.

Therefore, when the colonel had finished, no one accepted his invitation to stand forth and declare his opposition, and the meeting was dismissed with an order to load packs and proceed to the railway.

The next day, the fury of the Bolshevik offensive which swept the Vaga, and strove to realize Moscow's boast of annihilation for the Expedition, burst at Verst Four Forty Five where this "mutinous" company took the brunt of the attack and never wavered during the ceaseless, storming battles that followed, until, at the end of the third day, the enemy sullenly retired, repulsed and defeated, and another company relieved the exhausted American line.

And often before had these same men proved their mettle. There was no finer company in the regiment than this, and no more gallant officer than its commander. It is not the nature of the American to become "cannon fodder" without a question. Theirs was only the voice of sanity raised in this madman's war; yet when they saw that all in Russia were in the same plight, that no one knew the reason why, that all were caught in the same meshes of inextricable folly, they were soldiers, and played the soldier's part unfalteringly until the untried Russian conscripts came in May.

Many Russians had been killed as enemies; so like these simple peasants in soldier uniform that came to relieve the contested lines in May; so like the bearded host under whose foul-smelling roof the American dwelt. They did not seem soldiers; so spiritless, so immobile, so unmoved by firing emotions in this civil war wherein foreign defenders had died for Russia. If they felt any gratitude, it was covered beneath an exterior of impenetrable, Slavic lethargy, that defied all effort to disrobe. Life had been a thing of rote with these moujiks, as constant as the law of seasons and of stars, and the violent change from opaque darkness to the dazzling light, left them blinded, befuddled, groping for moral support. Before they had commenced to grasp the tremendous significance of the Revolution, swift came the Bolsheviks, crashing to earth every vestige of law, stability, the social structure, property rights.

Now followed these foreign invaders, warring upon the Bol-

sheviks and speaking with high sounding, noble phrases of saving
Russia, as they burned moujik homes and turned moujik women
and children out upon the cold snows. It was too much for the
poor serfs imagination. From fatalistic refuge he looked out on a
howling storm-tossed universe and abandoned all hope of comprehension.

Nitchevoo. There was no reason left on earth. All had gone crazy; all were stark, raving madmen in a madman's world!

So did the curtain fall on this lurid melodrama and its fretful
Railway scene, and now that the heyday of the fight was done, disquieting reflections took possession of the Americans. Their dead
had died for a scant few miles on this Railway battle ground, but
what the paltry little gain meant now not one could tell, nor why
the fearful price was paid, and ever came distracted thoughts of the
futility of it all, thoughts like howling, evil genie that ever recurred
to haunt and taunt those that came away.

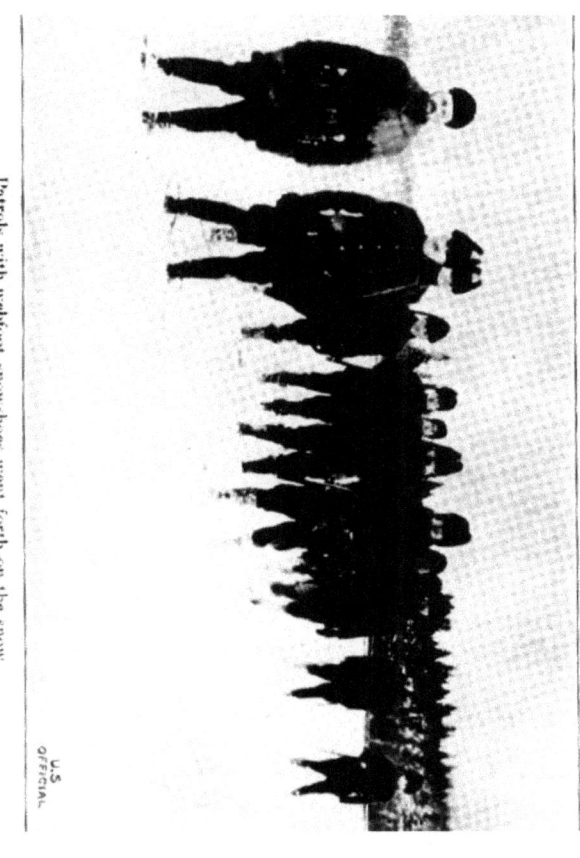

Patrols with webfoot snow shoes went forth on the snow

U.S. OFFICIAL

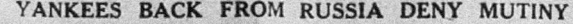

YANKEES BACK FROM RUSSIA DENY MUTINY

With steel helmets of white above their ruddy faces, forty-six officers and 1,495 men of the 339 Infantry, first American troops home from service in Northern Russia, arrived in Hoboken aboard the Von Steuben. The 339th is the regiment about which the "mutiny" story, which aroused nationwide comment, was circulated several months ago.

"Pure bunk" was the characterization of the story by Major J. Brooks Nichols, wealthy Detroit manufacturer and clubman, who was in command of the returning troops.

"More bunk has been published about the North Russia expedition than any phase of the war," said Major Nichols. "What gave rise to the story that Company I, of the regiment, had mutinied was an incident in which an order was misunderstood by a soldier who could not understand English well."

(1) Men of Company I, 339th Infantry. (2) Sergeant Matthew J. Gradok. (3) Sergeant Harvey Minteer. (4) Major J. Brooks Nichols. (5) Captain H. G. Winslow.

ONEGA

13th Feb., 1919

"Americally Sowest London for H.A.E.F. France. Due to prim- itive conditions of life and continuous service in the field under Arctic conditions, officers and men are beginning to feel the strain. Practically the whole Allied Command has been on continuous duty in the field all winter with no reserves in Archangel. Lim- ited Allied reserves are now being supplied from Murmansk, a few coming on ice breakers and others by rail to Kem and then by horses and sleighs to destination. Recommend present force be entirely replaced as early as practicable in the spring, with an adequate force commensurate with its mission, supplied and equipped so that it can operate in an American way."

-STEWART

Major-General Sir William E. Ironside

ONEGA

GENERAL IRONSIDE became Commander-in-Chief of the North Russian Expedition at the commencement of winter, and the "offensive war" forthwith came to an abrupt termination, without ceremony.

At that time, one company of Americans and ninety-three Slavic Legionaires composed the Onega or right wing of the Allied army which was at Chekuevo, some hundred and forty miles from Archangel on the Onega River.

A landing party of the original Poole force, expert rifle marines from the United States warship, Olympia, had taken the port of Onega after a noisy fight in September, and a few days later, gave it over to this Russo-American detachment, three hundred strong, whose object was to accord right lateral support to the Railway Column, and above all to safeguard the significant winter road connecting the Railway with Onega, along which the winter mail came sporadically, and the only reinforcements, three companies of British Yorks, were brought from Murmansk during the cold days of February.

As the Americans, verst post to verst post, fought their way south along the Railway line, so this detachment went forward at bloody experience and kept abreast, until the Bolsheviks, following the Railway victory at Verst Four Forty Five, grew cautious, and drew back up the Onega Valley to Turchasova.

And when winter came, the forty miles between Turchasova and Chekuevo, were a shadowy No Man's Acre along the twisting, snow highway of the river, where hostile patrols prowled, and life was held by uncertain tenure; but the disputed ground was narrowed by half when the Americans moved up part of their small number nearly midway to the Bolshevik village, and took station at Kyvalanda, in order to watch a southern trail inlet to the important Rail-way road, along which were regularly dispatched visiting patrols to the scattered villages of Bolshie Ozerki, that they might hearten and keep contact with the few pathetic Frenchmen and Allied Russians who made an audacious pretense of maintaining a post there, and far off on the snow, deserted many miles from the Railway, reminded one of a choice morsel of tenderloin, baited for puma.

The Onega detachment joined in the operation for Plesetskaya, which the new Commander-in-Chief, in furtherance of his defensive policy of consolidation, was anxious to take before the intense cold.

Plesetskaya was an important base, and had they lost it, the Bolsheviks would have encountered great, almost insurmountable obstacles, in bringing troops from Vologda, and concentrating them in an aggressive winter warfare, for this point was a junction of the principal highways leading from the Railway line to Onega, Kochmas, Tarasovo and Shenkurst.

But this Allied advance failed, primarily for the same cause that the whole Expedition failed, through ridiculous paucity of numbers, and in the second instance (although there were several more), because it was impossible to maintain any semblance of liaison over the difficult lateral terrain which separated the five Columns, theoretically converging in the push for Plesetskaya.

So on New Year's day, after they had met the enemy and soundly punished him in two sharp engagements, and standing to, were about to drive him from his Turchasova stronghold, the Onega Americans were given the disappointing order to fall back and resume post at Chekuevo, where long, black months followed, and life took on a grinding, monotonous, drab, depressing atmosphere, lifted only by an occasional, welcomed brush or "wind up," till lo, in March, the sun shone high and streamed in extravagant, effulgent light on the glaring snow fields, the days grew longer and still longer, in this eccentric, topsy-turvy, North world, and finally there were as few hours of darkness as there had been of light a few months before.

Late in the month, a patrol was driven off from Bolshie Ozerki by the shot from many rifles, and a combat party the next day ran counter machine gun emplacements, was extricated only by adroit leadership, and after worming a long distance through the piling drifts.

It was learned then that the little garrison at Bolshie Ozerki had been annihilated, but it was thought by a strong raiding party, bent upon capture of the ration and ammunition convoys between Onega and the Railway. Not yet was there a suspicion of the enemy's surprising, gigantic manoeuvre, which with incomparable, superior force, sought to turn the Allied flank at Obozerskaya, carry through to the Dvina, fuse with the Bolshevik Vaga army, then sweep on to Archangel and make good the Moscow boast to cast every foreigner in North Russia into the White Sea.

The British Colonel, irritated by the enemy resistance at Bolshie Ozerki, was determined to chastise "the raiders" thoroughly, and felt very confident when his seventy Americans were joined by the three companies of Murmansk Yorks, which had marched one hundred and seventy miles from Soroka on the Murman railway in the hope of reaching the hard pressed Vaga Column, before it was too late.

The only access to Bolshie Ozerki from the west is a wagon road, eighty feet wide, which cuts a swath through the ambient forest. Passing sleighs had packed this road so that it gave good going, but at either side among the trees was a hopeless, floundering snow bog nearly four feet, and two miles out from the village, the Bolsheviks had improvised an outguard, which swept this only approach with machine guns that had the concentrated fire of three battalions.

At dawn, on the twenty-fourth day of March, the Americans, supported by the Yorks on either flank, crept through the trees by the roadside to the attack on Bolshie Ozerki. At five hundred yards, the enemy opened fire, a murderous plunging storm of steel and lead that must completely quell all thought of further approach, still none turned back; dragging and pushing themselves through the snow by knees and feet and elbows, the men made four hundred yards; here the American officer was killed, two of the British officers were hit and went down as if struck by lightning, and it was seen by volume of the fire that the odds were hopeless, yet the little company, facing utter massacre, burrowed in the deep snow, and, in the stiffening cold, hung on to the last round, till the retirement order came at dusk; the sacrifice was a heavy one, but not in vain, for by this devoted stand the stupendous nature of the enemy operations to overwhelm the whole Expedition at Bolshie Ozerki was fully revealed, and every man at the rear position, vividly conscious of the desperate character of the fight, steeled himself for the grim business in hand.

Back in Archangel, General Ironside saw in a flash that the life of his army fluttered in the balance. He scoured the city for every available fighting man, collected the few he could, a varicolored assemblage of Americans, British, Allied Russians and a platoon of French mounted on skis—Le Legion Courier du Bois[31]—all counted, five hundred eighty men, and rushed with them to the battle. There, this iron General, well knowing himself to be faced by great unknown numbers, tossed caution high to the four winds. He dragged his artillery over the snow from the railway at Obozerskaya, and set it twelve miles off in the woods, daring the enemy to capture it. He brought out his handful of divergent troops, and, smashing down trees, built up rough barricades, a cordon about his guns; then, cut off from all hope of accessible retreat, this fighting man, whose fighting stuff had been welded among the Northwest Mounted Police of the Canadian frontier, threw down the challenge of wild death battle to the Slavs.

Very close, not even a mile away, down the Bolshie Ozerki trail, the Bolsheviks had concentrated their artillery and thrown out their advance works, and now commenced a blasting duel between the opposing batteries that tossed skyward mountainous geysers of snow, made fragments of the trees, and, through every lighted hour, shook the forest end to end with a ceaseless, reverberating roar, that pounded upon the ear with the vindictive echoes of tortured damned souls.

Fortune is a fickle mistress, but she loves the strong and smiles her favor on the brave, and in this strange mad Arctic forest fight, the Briton gained her countenance by thus handsomely risking all at a throw, and by his dashing courage, his magnificent, irresistible, reckless courage.

The Slav, more cautious, and overestimating the strength opposing him (as the Bolsheviks did time after time), did not strike while the iron was hot, but held off until he had gathered together three regiments; the 2nd Moscow, the 96th Saratov, the 2nd Kasan and several companies of ski troops; and the road that paralleled the Railway line to the Bolshevik camp at Shelaxa, near Plesetskaya, became a pitiful trailing havoc of dead and dying horses, driven to exhaustion in hysterical haste to bring still more artillery, more supplies, more ammunition to the waiting assault.

But every day spent by the Bolshevik chief, in fortifying his attack, was bringing victory to Ironside. In this winter campaign, with lack of transportation and dwelling quarters, it was always impossible to concentrate overmastering numbers of troops without costly postponement of the striking assault. The most troops that

31   *Coureur des bois*, runner of the woods, was one term used to describe the French fur traders who colonized North America in the 1600s and 1700s.

could be assembled were assembled by the Bolsheviks at the Vaga
and Bolshie Ozerki—probably eight thousand to ten thousand at
each place, and these were brought together with enormous labor,
incredible striving, heroic suffering in the cold, which plundered
the soldiers' strength, so that they were weakened by privation and
shaken by much exposure, and in the case of Bolshie Ozerki, came
to the fight too late.

So this battle that might have taken the life of the Allied North
Russian Expedition was lost, the fleeting opportunity for success
sped away when after the first fell stroke the precious element of
surprise was profligately squandered. And the Americans, bracing
themselves for the storm, fell to under the engineers, and working night and day, erected a citadel in the woods, strengthened
the barricades and actually finished two bullet proof blockhouses
before the first battle shock Immense stores of ammunition were
stacked high about the guns, and as the men labored, their confident enthusiasm grew; every soldier, under the stimulating, mes-
meric influence of his great chief, knew, with unwavering faith, that
the fight was won, grew impatient in the blood lust, and whetting
his bayonet, waited like a primitive savage, serene in the unshakable
conviction "that one Allied soldier was the equal of twenty Bolsheviks." So, in truth, he had to be in the battles of Bolshie Ozerki.

It was a tactical custom of the enemy to attack the front and
rear positions, sometimes he struck both simultaneously, but seldom the flanks. Therefore, General Ironside placed his Americans
forward and back, where the gun emplacements were, and then
stood poised for the onslaught. If the law of averages traversed
its orbital course, all might be well, but if the Bolsheviks forsook
their usual custom, these dispositions might well prove fatal; for
although the Yorkmen were scattered among them as bolsters, the
green, Russian, Archangel troops on the flank positions were as
yet untried, and the presumption was against them in the pending
death fight that would give no quarter.

But when the enemy came at last, on the seventh day, he came
just as the General had speculated he would come in an attack
on the rear guns; then in greater strength followed through at the
front barricades. The next dawn, at three thirty o'clock, the full fury
of the assault was uncovered, as three swaying rows of men hurled
themselves forward like swelling, tidal waves, and when this forward attack was at its climax, a wild horde stormed the rear.

In such an encounter, the great chance of success is in overwhelming the weaker adversary by sheer preponderance of numbers, to palsy his intelligence by bearing down on him with an awesome multitude, and before he has recovered, sweep him off his feet. But with these Americans, there was no such terror wrought hiatus, for the very intensity of the situation seemed to electrify their fiber, and fire their brains with the steady, blue flame of coordinated intelligence; under these overwhelming tidal attacks these fighting men were never so alert, never so keenly and appraisingly aware of every event, never so thoroughly mindful of every tense situation as it transpired; for they knew that piling cumbersomely through those bogging snow depths, the oncoming Bolsheviks were shackled nearly as effectually as if bound with ankle ropes, and they were acutely conscious of the verity, that in the circumstances, one steady man behind a bullet proof barricade, deliberately directing a functioning machine gun, had the weight of three hundred rifles.

So now it was a glorious thing to be in the blockhouses and the log barriers and to witness those human multitudes surge on, then slacken, and falter and fail and shrivel as they came, while machine guns swept them line to line, and flank to flank, and piled the dead and left crumpled, moaning heaps of men, where red, ugly blotches widened on the snow.

By noon, the fury of the storm had nearly subsided, the Commander of the Saratov Regiment, thinking his troops had won their ground, rode on his white horse nearly into the defenses and was shot down as he came, and from this time, the firing became desultory, except when some violent commissar drove small groups forward to be killed, or others, made desperate by despair, sneaked creeping out, and so were killed, and the rest lay flattened on the snow, not daring to go forward or back.

At nine, the sun went down upon the tumult of a bloody, gruesome day; it became cold again, and there followed dusky, unnatural silence, shattered occasionally by the rasping crack of snipers' shots, where in that night of horrors, the unfortunate Bolsheviks passed the acme of mortal misery. For if defeated, they returned to their own camp, death was waiting for them, and ahead were the remorseless Americans ready to shoot on sight, without stint of mercy. So, fairly caught between two fires, they lay out through the endless, black hours of terrible cold and frost, and gangrene took a greater toll than all the gunshot wounds.

Yet great as was the enemy distress, all knew that when the next day dawned, new forces would come up and press on to another determined assault, and it was to divert as many of these reinforcements as possible, that General Ironside ordered the Onega Detachment to move against Bolshie Ozerki from the west.

That same night, one of the York companies left the Onega Detachment and followed an unreconnoitred trail through the forest to strike again the hostile village from the north at daybreak; but long before dawn, became confused in the darkness and was hopelessly lost when the attack began on the road where another British company was to move against the village. A Polish company of Archangel volunteers, who were to execute a corresponding south flank movement, came from Chekuevo too late, so that the brunt of the fight fell upon the unsupported Yorks on the road.

Thirty minutes after the first faint light, dogs, tied to trees by the Bolsheviks, sighting the approaching front attack, gave boisterous, barking alarm, and, on the instant, the woods were made hideous with the rasping rattle of many machine guns. Many of the little band were hit in this first storm, but the rest kept on, dragging themselves through the yielding, four foot snow, while inches over their heads, the air howled hideously with the passage of flying death. In the snow, rifles became clogged in the breeches, so that the bolts would not drive home, and men had to dig them clean with fingers stiffening from cold, but still, a little at a time, the attack wormed on and on. At one hundred yards, the gallant, British captain rose to lead a rush at the machine gun positions and was killed in his tracks; then the second officer was hard hit, and when the delayed Polish company came forward in support, and two of its number got shot through the bowels, the others bolted like sheep and could not be driven to the battle again.

Then the Yanks went in and stood manfully to the fight by the side of their distressed comrades, but against heart sickening, desperate, despairing odds, as the merging Bolsheviks came from both sides and massed in a vicious, determined counter attack that would have overcome all, but when doom seemed certain, the lost York Company emerged from the woods, by some act of a benevolent Providence, to meet and stay the fullness of the thrust, until darkness came down to save the valiant, little band on the snows.

This last, noble effort of the Onega Detachment had been made with a single thought—that of baring their breasts to the blow that otherwise would have fallen on their tired comrades in the barricades out in the forest from Obozerskaya; and great as the cost, its effect had been the final discouragement to the Bolsheviks who made one more ineffectual effort to gain the Allied Railway flank, then drew back in full retirement to the south.

The enemy sustained great losses in these battles of Bolshie Ozerki, upwards of two thousand casualties, many of them from the frost, for the villages could shelter but a fraction of the large forces, and many had to live in such makeshift quarters as could be devised.

Time was of the essence in this undertaking of the Bolshevik commander, and he had paused when he should have struck out with every man in his control, but by his dalliance, spring joined the league of his enemies. Soon the freezing clutch of winter would be broken in the warm sun, and, unless he hastened to withdraw to the south, his artillery would be mired in the yielding roads.

In June, the new, conscripted, Russian soldiers came to take Onega's posts, and the heavily-tired Americans went back to assembly at Archangel, buoyant and bright-eyed at the prospect of home, till they met on the city streets a few invalided Category B Scots going back to the battle lines, because the Bolo droves were gathering again and every man was needed there. Then the light smile passed from the lips of the Americans, a blush came to their cheek, home was forgotten and all thought of home; for there was a man's work out in the forest swamps far to the south—where death lurked and misery waited; and hardly a man who would not have chosen the swamps with their physical suffering and their ambushed death than escape and bear the stinging reproach of deserting a mate in distress. Better to play the wretched game through to the uttermost end than to be faithless to the traditions of one's blood, to quit the field with the honor of a nation stained and shamed in burning disgrace.

For was this such a flagitious, disgraceful brawl in which their mates had bled their manhood blood away that American soldiers should sneak from it thus, like cuffed and beaten mongrel curs?

Time, soothing time, will smooth with gentle, cooling fingers, the harsh lines of fretful hardship, the distressful burdens of campaign and trying vigils of sleepless peril, and even burn a purple halo of romance about this miserable, petty, little war, but some hurts the assuaging balm of time can never heal.

Many had cast off at the call of country and given all with generous unstinting affection, and those who were coming back did not begrudge the sacrifice; but rankling deep forever in the living consciousness of every Archangel soldier is the thought of this ignoble quitting and the weak abandonment by his country of everything to which he had pledged his manhood faith.—The causelessness of it all—Alarming, unbalancing reflections, a moral devastation that will not be quieted—Corroding grief for those who flushed with promise were "taken from life when life and love were young"[32] in a shabby brawl for nothing.—A dangerous cynical bitterness is with the soldier of North Russia, mordant and enduring, that grows ever more bitter with the years.

---

32   *The Grave of Keats* by Oscar Wilde (1854–1900)
Rid of the world's injustice, and his pain,
    He rests at last beneath God's veil of blue:
    Taken from life when life and love were new
The youngest of the martyrs here is lain,
Fair as Sebastian, and as early slain.
    No cypress shades his grave, no funeral yew,
    But gentle violets weeping with the dew
Weave on his bones an ever-blossoming chain.
O proudest heart that broke for misery!
    O sweetest lips since those of Mitylene!
    O poet-painter of our English Land!
Thy name was writ in water—it shall stand:
    And tears like mine will keep thy memory green,
    As Isabella did her Basil-tree.

ROME

The blockhouses where men were crippled and maimed and shell-shocked,
far away from gala Archangel

one-knit socks strung up to dry over camp fire at the edge of a
orest, through which the men of Company M, 339th Inf, pick their
ay for 17 hours, in a vain attempt to outflank the Bolsheviks farther
down the railroad. A great part of the march was through mud and
swamp. Obozerskaya, Russia. Sept.29,1918.

January 30, 1919.

MEMORANDUM FOR COLONEL HOUSE.

SUBJECT: Withdrawal of American troops from Archangel.

Dear Colonel House:

The 12,000 American, British and French troops at Archangel are no longer serving any useful purpose. Only 3,000 Russians have rallied around this force.

Furthermore, they are in considerable danger of destruction by the Bolsheviki.

The appended memorandum and map which General Churchill has prepared show that unless the ice in the White Sea suddenly becomes thicker, it is possible at present, with the aid of six icebreakers, which are now at Archangel, to move these troops by water to Kem on the Murmansk Railway, whence they may be carried by train to Murmansk.

The situation at Archangel is most serious for the soldiers, but it is also serious for the Governments which seem to have abandoned them. Unless they are saved by prompt action, we shall have another Gallipoli.

Very respectfully yours,

WILLIAM C. BULLITT.

Abridgment of communication from William C. Bullitt of the American State Department, delivered to Colonel E. M. House at the Paris Peace Conference, on 30th January, 1919.

KODISH

KODISH was the epitome of North Russia. Bought with toiling effort, incredible privation and cruel losses, to be lost and won again time following time in the bitter- most winter days with moving heroism and a moral grandeur that at times reached a sublime estate—it was in the end abandoned as "of no especial military significance."

The village lay in the course of the Imperial road from Petrograd that parted from the Vologda railway at Plesetskaya and cut a diagonal lane through the province northeasterly to Emetskoe on the Dvina. Both Commands stressed its importance. In the early days of the campaign the Allied leaders, bent upon conquest, seized upon it as an opportune route to support the railway invasion by surprising the enemy in rear, while the Bolshevik Staff saw a chance to drive a wedge between the two advancing Columns and effectually deny the River forces all communications.

A typical polyglot group of French, British, friendly Russians, and a few American marines, some two hundred in all, had gone out from Archangel in the first days of the Expedition to Seletskoe on the Emtsa river determined to drive south from this subsidiary base along this Petrograd road to Plesetskaya. This group, designated "D Force" to distinguish it from "A Force" on the Railway and "C Force" on the Dvina, and the Vaga, had hardly commenced its daring operation when an urgent call for succor caused British, French and Americans to hurry across a trail through the swamps to Obozerskaya, leaving the loyal Russians as rear guard before Kodish. But the former never reached their goal. Days passed and nothing was heard from them until a relief contingent, out a day's journey from the Railway front in the forest swamps, found in the midst of scattered infantry gear and other signs of desperate encounter the soiled diary of an American sailor with the epitaph of this illfated "B Force" written on 30th August.

The rescue party continued east through the swamps to Seletskoe as the pursuing Bolsheviks closed in on that village, but the Americans, reinforced by a slender garrison, drove them south over the Emtsa, where they stood their ground behind a destroyed bridge. It was suicidal to attempt a passage of the open river in the face of machine guns, so the Americans dug in the cold sodden ground, and in the grim siege that followed the suffering was intense; no doctor was at hand to care for the many casualties who were given crude first aid (when they were reached), and bumped and jolted thirty torturing miles to Seletskoe, yet, in the face of all these things, none at Kodish knew thought of weakening or turning back.

On the ninth day, long awaited supports came up, a crossing was effected at an unexpected point below the Bolshevik position, and Kodish succumbed to a courage that would not be denied. Exposed baldly in a broad clearing and flanked by three dominating hills, this moujik village was helpless against modern artillery. The French colonel pronounced it "strategically untenable," but the worst feature was its opportunity for complete encirclement. This was brought vividly to the consciousness of the Americans soon after their occupation when great Bolshevik bands converged on them from villages to the south and the Shred-Mekrenga trail, and following a four days' battle, they fought for their lives in a night flight nearly two miles along the road back across the river.

There the old familiar siege tactics were resumed. The engineers with a genius of adaptitude built a fortress of blockhouses on the north Emtsa embankment, and in these, one company of Americans, augmented by a few British infantry and a section of Canadian Field Artillery, stood off the Bolsheviks from the crucial Petrograd road. In December, with Plesetskaya the objective of three Allied fronts, this little group, now 450 strong, led by the impetuous "Major Mike" Donaghue forced twenty-seven hundred Bolsheviks out of Kodish, but could make little progress on the road beyond. So the contested village was held as an advanced post for the main Allied force on the Emtsa, and exposed to unremitting bombardment from many superior guns, became an inferno of bursting shells.

Once on a black January night, it was abandoned by the little outpost and set aflame, but before dawn, Donaghue was back with his men to a chaos of charred ruins, like the skeleton of a beast of prey in a desert of snow, through which the bitter, chill winds wailed dolefully. In these deserted Kodish streets of abject desolation, the American soldier knew the uttermost depths of physical misery experienced during the whole winter campaign.

The Commander-in-Chief came to the Kodish front when British soldiers evinced a truant disposition and would not "carry on" unless certain interrogatories concerning this evasive war with

Russia were answered. The interrogatories were addressed to Premier Lloyd George and were such as might arise from the mental consciousness of any men who still have well poised, wholesome regard for life and the pursuit of happiness as they understand it. These British soldiers had come from the winter murk of Murmansk, had emerged from four years' hell in France, and saw themselves the hapless forfeit in a confused international melee without wit or reason at a time when all were thoroughly sickened with war and thought they merited restoration to their homes. But when the soldier Ironside,[33] six feet four, with "an eye like Mars to threaten and command"[34] had spoken, the interrogatories were all forgotten and these disgruntled men, who had uttered mutiny, returned to the fight with a matchless valor; with a steadfastness that gave never ceasing wonderment that they could so freely offer all with every instinct and inclination opposed.

It was at Kodish that the Bolsheviks strove their uttermost with propaganda, that insidious, warring weapon of which so often they have revealed themselves the masters. Thousands and thousands of pamphlets, leaflets, circulars, manifestoes, announcements, proclamations, appeals—an amazing collection of vitriolic, eloquent literature, were left along the patrol routes in the snow forests. This was true at all fronts, but especially at Kodish, where these persuasive methods were concentrated like a great verbal bombardment, a veritable war of scarifying words, Russian, French, German and English. Many messages of hate and fire, with frank artlessness, urged the Allied soldier to desert and join the Soviet; others, more subtle, displayed a masterful knowledge of human weakness and human passions and prejudices.

The following is taken from *The Call* published in Moscow and printed in English:

> *Do you British working men know what your capitalists expect you to do about the war? They expect you to go home and pay in taxes figured into the price of your food and clothing, eight thousand millions of English pounds or forty thousand millions of American*

---

33   Field Marshal William Edmund Ironside (1880-1959) was an artilleryman in the Second Boer War, and served on the western front for two years. After North Russia, he was sent to Turkey and Persia. He briefly served in WW2.

34   Hamlet, Act 3, scene 4, speaking of his father: "*See, what a grace was seated on this brow;/ Hyperion's curls; the front of Jove himself;/ An eye like Mars, to threaten and command;/ A station like the herald Mercury/ New-lighted on a heaven-kissing hill;/ A combination and a form indeed,/ Where every god did seem to set his seal,/ To give the world assurance of a man:*" (55-62)

dollars. If you have any manhood, don't you think it would be fair to call all these debts off? If you think this is fair, then join the Russian Bolsheviks in repudiating all war debts. . . .

Do you realize that the principal reason the British-American financiers have sent you to fight us for, is because we were sensible enough and courageous enough to repudiate the war debts of the bloody, corrupt old Tsar? . . .

You soldiers are fighting on the side of the employers against us, the working people of Russia. All this talk about intervention to "save" Russia amounts to this, that the capitalists of your countries are trying to take back from us what we won from their fellow capitalists in Russia. Can't you realize that this is the same war that you have been carrying on in England and America against the master class? You hold the rifles, you work the guns to shoot us with, and you are playing the contemptible part of the scab. Comrade, don't do it!

You are kidding yourself that you are fighting for your country. The capitalist class places arms in your hands. Let the workers cease using these weapons against each other, and turn them cm their sweaters. The capitalists themselves have given you the means to overthrow them.

...if you had but the sense and the courage to use them. There is only one thing that you can do: Arrest your officers. Send a committee of your common soldiers to meet our own workingmen, and find out yourselves what we stand for.

The following is from the same publication:

The Bolshevik Revolution marked the culmination of the world struggle to set us all free. Strike off your shackles, comrades, we are your friends not enemies, and the only reason we seek to stamp out the parasitical capitalists by force is because force is the only language they can understand. This is the beginning of a great world revolution which knows no national limitations. It will set the producers free. Join the Soviet Party. We are fighting your fight against the unprincipled capitalistic class. Comrades, you know the meaning of "scab," well, that is the part you are acting in Russia. For shame, comrades! Kill your officers, then shoulder your rifles and come over to our lines which are your own.

These extracts have been taken at random from a hundred others of like incendiary tenor, most of which had little effect on the Americans except to impress them with the coincidence of a striking similarity in style and sentiment between them and many public addresses of American politicians printed in the newspapers that came from home, where a soft going government tolerated perversions of free speech, as hostile to American soldiers in Russia as the most violent preachments from the enemy.

A huge bulletin board was erected on the Bolshevik bank of the Emtsa river, which conducted daily classes in doctrines of International Revolution, and the first confirmation of the Armistice news came in a weird preternatural voice which startled the night stillness of Kodish by announcing in sonorous tones the cessation of infamous war and the restoration of peace to the afflicted peoples of earth. There on the Emtsa bridge, a Bolshevik orator, shrouded by the phantom shadow of a waning moon, delivered in excellent English, almost academic in polish, a rhetorical harangue on the glories of communism, the injustice of soldiers suffering in cold swamps while others sat back in Archangel in soft ease. Also the speaker described most persuasively the abundant, bountiful hospitality awaiting all within the Soviet lines. It was all very diverting, but nevertheless gave audible utterance to many of the disquieting reflections which rankled deep in the heart of every man in the Allied ranks and did not go towards helping Allied morale. Later that same night, when this extraordinary speech was ended, two captives, a Scot and an American, came out on the bridge to tell their comrades of benevolent treatment at the hands of the unspeakable enemy; in the darkness their voices were like those from the grave, for many soldiers were led to believe that the barbarous Bolos killed all prisoners after torturing them with frightful savagery.

In the first stages of the campaign, the French on the Railway killed those that could not be carried off the field to spare them the grewsome horrors which would have been visited upon them by the enemy, yet at Ust Padenga, volunteers brought in wounded not a hundred yards in front of Bolshevik machine guns, and at Toulgas, after a disastrous ambush, the enemy mysteriously withheld his fire from a relief party that was entirely exposed. There was, in fact, only one recorded instance of atrocity. This was on the Vaga where the bodies of an officer and several Americans were found hacked and mutilated with hideous debauchery, but there was nothing to show that this barbarism was approved by the Bolshevik leaders, and it may have been only an uncontrollable manifestation of primal cruelty which underlies all war.

Several months after the last troops left Archangel, a number of
Americans "missing in action" were expatriated through the efforts
of the Red Cross by way of Finland, and these men spoke very
favorably of their considerate treatment in Moscow.

Temporary Red Cross Hospital. September 21, 1918.

THE RIVER

"There ought to be an efficient American Hell Raiser from one end of the front to the base, with a rank of lieutenant colonel."

DOCKS JOHN HAU. (Major Medico 339th U. S. Infantry). 21st October, 1918

"The Government of the United States has never recognized the Bolshevik authorities and does not consider that its effort to safeguard supplies at Archangel or to help the Czechs in Siberia have created a state of war with the Bolsheviki."

Cablegram, State Department, Washington, D. C, to David R. Francia, American Ambassador, Archangel, Russia. 27th September, 1918.

THE RIVER

HALF of the original Poole Expedition was selected for the punitive pursuit down the railway, a garrison was left to guard Archangel, and the trifling group that remained followed the dark course of the Dvina into the unknown region of the interior. There were told off for this river expedition two depleted companies of the Tenth Royal Scots Regiment, and twenty-five of the American marines crowded into merchant barges and towed slowly up-stream by small tugs. The only escort was an armored British monitor, and seen from the shore, as they made their toilsome struggling way against the swift raring river course, conspicuous, unshielded targets on its broad surface, the dauntless little band looked tempting ambush prey.

At Chamova, some one hundred and eighty miles from Archangel, the enemy gave sign of having abruptly recovered from his first stampede. He turned and showed his fangs, and the pursuit stopped short.

It now grew apparent that the retreat had not been as riotous as first supposed; in fact, there was good reason to believe that it was a part of Bolshevik strategy, and evidence was accumulating that Trotsky had ordered the withdrawal from Archangel to make certain of the millions of American made supplies and ammunition, had taken a careful appraisal of the military situation, and elected to give battle in the interior. When the Americans arrived they were met at the wharf by an agitated Brass Hat who said the Allies at both fronts were standing at bay and the situation had assumed a very precarious phase.

The Third Battalion was rushed to the Railway, and the First Battalion, in dirty, ill-smelling barges, followed the pioneer Poole Expedition up the river one hundred and fifty miles to Bereznik. These barges had carried many cargos on Dvina's waters, cargos of livestock and flax and other agricultural produce, but were new to human freight, and in their cramped, miserable, dank quarters, the scourging influenza broke out afresh among the troops, and those who had already been weakened by the disease grew fainter and fainter as they followed up the unknown waterway till a day came when one after one they quietly passed to the bourne of that country of gentle unwaking sleep, and sometimes off on the gloomy foreboding river the passage of this antic caravel seemed more a funeral processional than an aggressive expedition of war.

The tired comrades who were even denied the vibrant thrill of the fight, and its doubtful glory, were with simple soldier ceremonials given to the soil of Russia, ceremonials, moving because of their simplicity and that wholesome, fullhearted sentimentalism which has always marked the American character—and always must be of our America.

Here in these little churchyards, tragic death seemed robed in sorrow more sacred with the brown, barren embankments like a shroud of mourning, the grave skies drooping and disconsolate and the sombre recesses of the forest where taps trailed in grieving cadences and echoed within the soldier's spirit long after its last note had been lost in the gloom. Laden with inarticulate depression and confused melancholy, thoughts of life's crazy theatre, the crushing power and immensity of fate, the tragedy of all, these men fresh from the fields and shops of Michigan and Wisconsin groped their dazed way back to the barges where dark shadows with ominous fingers reached over the waters and death, in this haunting, melodramatic land waited, suspended in the alien air like a pestilential vapor.

The first stop was five days out from Archangel at Bereznik, near the junction of the Dvina and its main tributary, the Vaga. Here there was a group of commodious, well constructed log buildings, which had served as hunting lodges for the Tsar Nicholas and his retinue during the days of the Romanoff dynasty. It was decided to make use of these buildings for storage purposes, and to have Bereznik as the subsidiary base of the Dvina expedition until progress was made so far up the river that practical considerations would impel the movement of the subsidiary base to a more advanced position.

So from the time of the arrival of the Americans on the 13th September, until the close of water at the end of October, rations, munitions, clothing and other accoutrements of war, in value over one million pounds sterling, which had been brought all the way from England, were loaded on every craft that could be commandeered at Archangel and transported the one hundred and fifty miles to Bereznik.

One of the American companies was left to guard these precious supplies and the others hurried on to take up the gage of offensive campaign. There was a brush at Chamova, but the enemy did not make his first stand until he came to Seltzo, nearly thirty miles further upstream, and now well over two hundred miles from far away Archangel. Except on the Vaga, this was the furthermost south achieved by the Allied troops.

At Seltzo, it became clear that the Soviets had no intention of running further, and that the foreigners would be fortunate if they held the ground already gained. The tactical abandonment of Archangel having accomplished the effective seizure and retention of everything of value in that port and extended the invader far into the interior, revealing with obliging frankness his numerical weakness, had realized the ends sought by the Bolsheviks, and the signs were now many that they intended to strike back and strike back hard.

Why did not Poole retire to Archangel?

The futility of the attempt to reach the distant Siberian railway with the ridiculously small force at the disposal of the Allied Commander was glaringly apparent to every common soldier.

Why did not Poole, like Joffre at the Marne, shift his policy to meet the exigencies of the military situation, draw in his far scattered fronts to Archangel, construct an enceinte of defenses about the city, and hold on until help came in the spring, or until some definite action was determined for Russia?

An outpost on the Railway

Many lives would have been spared and much misery averted
had this been done, but the lives of a few men, and the permanent
impairment of the lives of many more, do not weigh heavily in the
scales with those who sit in the councils of the inner sanctum at
General Headquarters and think nothing of the spending of divisions and even army corps. Perhaps it would have been too galling
to Anglo-Saxon pride to admit being on the defensive before an
inferior people like these poor Slavs who were to be chastised with
thoroughness and dispatch. Then, too, it was always safer for Archangel to have the outposts far into the country, and flattered the
Allied Command in the belief of still being the aggressor.

When Ironside took command he not only conceded that the
Allies were conducting a defensive campaign, but with soldier
bluntness declared that the Expedition was in gravest peril. It was
too late then to draw in the far dispersed battalions. They would
have to fight it out on the wide separated snowbound fronts, and
show by deeds the superiority of the Anglo-Saxon. If they failed,
if they were faint hearted and even so much as faltered, the entire
force was doomed.

On the morning of 19th September battle was joined at Seltzo.
A mile of open marsh lies outside, through which the stream at the
border of the village meanders from the forest to pay tribute to the
mighty Dvina. The only easy approach is along a narrow road that
parallels the river and crosses a bridge over this deep icy stream.
On this morning of battle the Americans waded the swamp until
within fifteen hundred yards, when suddenly from the scattered
concealment of the houses there burst such a furious fusillade of
musketry and machine guns and Pom Pom guns that they dropped
low in their tracks and could go no further.

Two other companies moved through the woods on the flank
to assist the frontal attack, but their location was determined by
the enemy batteries, and his infantry laid down such a withering
fire, that the battalion, exhausted from a day of fighting and a
heart-breaking march, without rations and with no cover from the
cold and the drizzling rain, was compelled to bivouac that night in
the soaking morass, hopeful that with next morning would come
promised artillery support, for without it further advance was un-
thinkable.

All through the night the Bolshevik guns searched for the
Americans who were new to combat, ignorant of the ground, and

had not an inkling of the enemy strength or his fortifications or dispositions. And at dawn a recon-naissance patrol stumbled into a large enemy force, was scattered and came back with no information, save that the Bolsheviks had assembled in superior numbers and were well supplied with ammunition. As daylight broadened, the shelling from the river became so violent that the attackers had to choose between a further advance or complete retirement; to stay where they were meant destruction.

So with grave misgivings the attack was renewed, although there was still no sign of promised artillery support; machine guns guarding a trench system in the woods killed and wounded many Americans, but the advance would not give ground, and supporting comrades at flank and rear kept up such a sustained unfailing fire that the Bolsheviks were led to believe that the attack had been replenished during the night.

During the fight the American lieutenant colonel "caught in a bracket" had stayed in the rearward village, Yakovlevskaya, but at dusk he emerged with the important Field Pieces which laid down an effective feu de barrage on Seltzo. Hardly had it lifted when the battalion arose and with splendid dash and gallantry stormed forward to the village, entered it and took possession. But the story of Seltzo is the story of the whole campaign. After the infantry, with inspiring display of courage and at great cost, had gained a position, its small forces would be drafted for some other distant hard-pressed front, or the position would be left to the mercy of the Bolshevik guns until no course was left except evacuation.

The monitor which had convoyed the battalion up the Dvina, fearful of being caught by the ice that was expected to creep upstream from Archangel at the beginning of October, but did not actually come until mid-November, went back before the battle and was gone for the duration of the winter. A few days after the battle, the artillery left and was seen no more at Seltzo. Also Headquarters ordered two of the companies to proceed to Shenkurst on the Vaga, the second city in the Province, where it was alleged a large number of Russians in sympathy with the Allied cause were anxious to have a garrison of American troops during the approaching winter.

So it came that there was no artillery to avenge the smashing havoc of the enemy heavy guns in this furthermost Dvina village where one infantry company of Scots, a like number of Americans, and a few Allied Russians held on under terrific shell fire that from river and forest racked and battered them.

The enemy had a complete battery of three inch pieces, which he was free to bring up to the edge of the woods beyond the village, and down the river on rafts and improvised gunboats he floated three six inch guns and two Nine Point Two naval pieces, and for days with this combined armament he smashed and blasted until many of the houses became a riot of shredded and splintered timbers, and it was only a question of time before the garrison would be decimated utterly.

On 14th October the Bolsheviks attacked the defensive positions with great vigor, but were thrown back in complete repulse with many killed; yet that night and in the first morning hours the defenders slipped away in the darkness, for under unhindered bombardment the place had become a death's trap where all must eventually perish.

After this escape in the night there was a heart-breaking drag through the mud, until late the next day the tired Allied soldiers found harbor in Toulgas some fifteen miles back. Toulgas is typical of the North Russian village, a group of bedraggled log houses huddled together on a hill, which bends down in a long easy slope to the plain, where, like Seltzo, a stream comes out of the forest and margins another cluster of huts on the flat ground which the moujiks call Upper Toulgas.

This stream is deep and numbingly cold, and has cut an abrupt channel through the yielding soil so that fording it is a difficult feat at best. For an enemy to make the attempt in daylight would be suicidal. In darkness, any considerable numbers cannot fail to give the alarm. A road comes down from the hill and crosses a wooden bridge to the forward village. Watching the bridge is the inevitable white church, and its gaudy minarets, consciously aloof and superior in the poverty of the scene. In the setting of dun barren ground the white edifice flashes in undefiled purity against a low shrouding sky, more black than gray, which rests upon the darker tufted forest.

U.S
OFFICIAL

The fighting Canadians

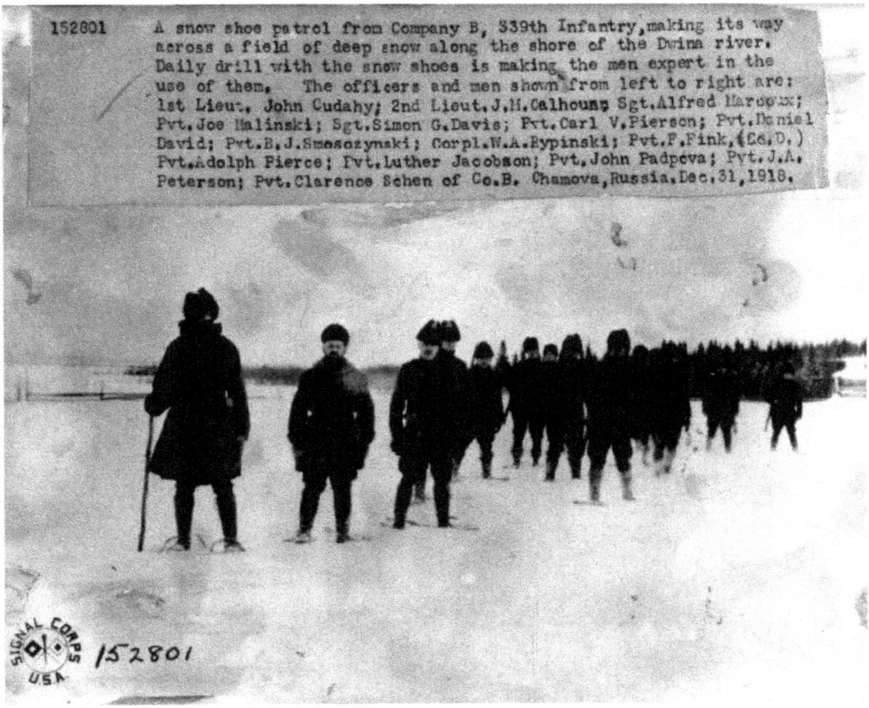

152801   A snow shoe patrol from Company B, 339th Infantry, making its way
         across a field of deep snow along the shore of the Dwina river.
         Daily drill with the snow shoes is making the men expert in the
         use of them.   The officers and men shown from left to right are:
         1st Lieut. John Cudahy; 2nd Lieut. J.M. Calhoun; Sgt. Alfred Maroux;
         Pvt. Joe Malinski; Sgt. Simon G. Davis; Pvt. Carl V. Pierson; Pvt. Daniel
         David; Pvt. B.J. Smasczynski; Corpl. W.A. Rypinski; Pvt. F. Fink, (Co. D.)
         Pvt. Adolph Pierce; Pvt. Luther Jacobson; Pvt. John Padpova; Pvt. J.A.
         Peterson; Pvt. Clarence Schen of Co. B. Chamova, Russia, Dec. 31, 1918.

152801

Across the road is the priest's house, like the others of bark stripped logs, differing from the others only in its greater size. With a little barricading the walls of the priest's house were secure against the lead of small guns, but it was death to stay there during the avalanche of high explosive shells that was poured out by the Bolshevik gunboats.

After the Battle of Armistice Day, the bearded priest of Toulgas Church was found amid the hideous battle litter of his wrecked home, the crown of his head cut clean as with a scalpel, exposing the naked brains. Near him were two children, a boy and a girl, sleeping by the guardian who from infancy had taught them of a Providence who watched over the good of earth, and surely would not desert them through this malignant turmoil that had descended to the quiet moujik country with terrible death and indescribable misery like the recurrent plagues. So sleeping, a shell had found the unconscious children, and lulled them to that everlasting sleep. The big shells had a way thus, of stealthily sniping their victim's life away with no mark of their dread approach, as if disdaining the brutality of violence. But again they would pounce down with the atrocity of a fiend, smash head from trunk, and members from the torso, and leave great gaping wounds gushing black blood with unspeakable, horrible ghastliness.

Back of the church, on the same side of the road, is a moujik house with the customary stable attached in rear. A platoon used this as billeting quarters. It was shielded by the church forward, and gave shelter to the little reserve that would replenish the blockhouse at the bridge with men and ammunition, and, if the blockhouse was knocked out, would stand off the Bolsheviks from crossing the bridge.

From the billet house to the church is about thirty yards. The priest's house is nearly opposite the church across the road. The blockhouse was built just before the Armistice fight and stands on the bank of the stream guarding the bridge about twenty yards forward of the priest's house. It is thirty yards over the bridge, and in front of the first line of Upper Toulgas houses, a field, shorn of all cover, stretches one hundred yards to the stream.

Back of the center village on the hilltop the ground undulates almost unnoticeably in a series of folds and reaches a shallow draw. A little beyond this, perhaps two hundred fifty yards, is still another clump of huts known as Lower Toulgas. In this draw, the Canadians built emplacements for their two Field Pieces, which during the first battles were the only artillery for the defense of Toulgas.

The forest gives way for nearly a half a mile before Upper Toulgas. From Upper Toulgas to Lower Toulgas is an ample two miles. From Toulgas, itself, the center village, to Lower Toulgas is a scant three-quarters of a mile.

On the forest flank the ground has been cleared for a space, varying from three hundred to less than sixty yards. This clearance is greatest opposite the upper village. In the lower village it narrows, until in rear the trees close in on the road that leads back to Bereznik and Archangel, affording excellent opportunity of concealment and surprise attack for an enemy that would have the endurance and the hardihood and the courageous daring to march through the deep swamps of the woods.

On the left the Dvina spreads out in a wide expanse, two miles. Opposite the rear and center villages the river banks are high and steep, nearly precipitous, but at the forward village on the flat ground the level is only a few feet above that of the water. Across the river there is not the slightest sign of cover as far as the distant embankment on the opposite shore. The chances for surprise from this quarter are practically none, and without surprise, infantry advancing over the waist-deep snow against machine guns, would have to be possessed of fanatical courage and be in overwhelming strength. The river could be nearly neglected as a source of danger.

To defend the three Toulgas villages we had: One company of American infantry; one company of Royal Scots infantry, and one section of Field Artillery, manned by fifty-seven Canadians.

In command of this force was Robert P. Boyd, an American civilian, who, scarcely a year before, had graduated with the rank of captain of infantry from a three months' officers' training school at Fort Sheridan, Illinois.

Shortly after occupation of Toulgas, ice choked off navigation of the lower river, and replenishments of supplies and ammunition had to be brought by small one-pony sleighs from Bereznik. The distance was some fifty miles, and the journey by Russian pony was usually two days, but when the snow was deepest, the weather bitterly cold, and the days had but few hours of light, it took three days.

There was a field hospital at Bereznik, vicariously supplied, and attended by a medical personnel of changing nationality, British, Russian and American by turns.

We converted one of the huts of Lower Toulgas village into a dressing station, where first aid was given the wounded; but we had no facilities, no operating equipment, or surgeons, or surgical instruments to care for the serious cases. If a soldier was hard hit and lived, he had to be brought to Bereznik.

Following the retreat down river from Seltzo, there was hardly time for a tactical survey of the situation, for the construction of temporary redoubts on the forest flank and at the crucial bridge, when enemy gunboats opened fire on our positions and for three days kept up a determined bombardment. When dusk came on the third day, the shelling lifted, and when the night grew black there was a roar of many rifles and a mad yelling from the woods as a horde of Bolsheviks fell on the center village. In the darkness and wild confusion, the tumult of battle made by the roar of musketry, the shouting and screaming of many foreign voices sounded like the onslaught of a Division.

But, even with the advantage of overpowering numbers, a night attack to succeed, demands most accurate knowledge of the enemy position, and most rigid control by a leader of his men. The Bolsheviks were not thoroughly trained in these early days, although later they displayed impressive military skill and the utmost cooperation between officers and men; now their lead went high and shrieked through air several feet above the heads of the unscathed Americans, who had concealed Lewis guns in a dugout at the point of the enemy rush and turned these loose upon the massed Bolsheviks, felling them like cattle in a slaughter pen. One American private, swinging an automatic rifle from his hip, shot until there was a semicircle of prostrate forms before him, some of them fifteen yards away; and once a few of the enemy came so close that they were spitted at the end of the bayonet.

At the height of the fight the Canadians opened up their guns and rained the woods with shrapnel which threw the wavering Bolsheviks into worse commotion and disorder, for while the Lewis guns scattered death in front, rattling shrapnel bullets threatened death in rear, and thus, huddled together in the darkness like stampeded sheep, they were shot down until the fierce exulting battle yells were changed to moans of the wounded and appealing cries for mercy.

At a signal, the Canadian guns ceased firing, the Royal Scots, shooting low and true, went into the counter, and the disorganized Bolsheviks, seized with blind animal terror, lost all semblance of order and fled in violent flight, each man for himself, to the sheltering recesses of the forest.

After this night attack there was nearly a fortnight of quiet on the Dvina, with no outward sign to show the enemy intentions. Patrols went out into the woods and came back with the report that Zastrovia, the nearest village upstream, was clear of hostile troops; but, while the Allied Command took under advisement the opposing contentions of retirement and holding on, the Bolsheviks were assembling large fresh forces of infantry, and bringing heavy guns from Krasnoborsk, preparatory to striking the most ambitious blow yet attempted.

All at Toulgas were aware that the lull was ominous. All knew that this phase of security was a very transient one, and directed by the American engineers, every man who was not on guard duty, worked building log blockhouses, at tactical strong points about the center village, one of them to guard the bridge over the stream to the upper village, where there was a small outpost, which in case of frontal attack was to give the alarm, then retire to the defenses.

The defense centered around the middle village. There were no fortifications to protect Lower Toulgas, and the Canadians in the draw in front of Lower Toulgas had for their protection only a squad of Americans under a sergeant, with a Lewis gun. The great danger in the situation lay in the threat of the capture of the rear village by an attack from the close-edging forest. If this lower position was taken, the garrison would be trapped, starved and cut off from all communication with Bereznik and Archangel. Customarily, there were kept on hand rations sufficient to last from two to three weeks.

When the British Brigadier General R. G. Finlayson inspected the Toulgas area, on loth November, apprehension of such a rear attack was expressed by some of the officers, but the general could see no real menace from that quarter, and said that it was a military impossibility for a large body of troops to successfully execute a flank movement through the heavy swamps of the woods.

The day following, Armistice Day, at dawn there was a crackling of rifles in Upper Toulgas, then the crash of guns from the river, as a great number of Bolsheviks swarmed from the forest, deployed in perfect order, and advancing in squad rushes, drove the little outpost back to our main lines. Timed, it seemed almost to the moment, came the roar of musketry far at rear, the staccato rattle of machine guns and dominating all the din and tumult, the ringing Cossack Hourra! Hourra!

Our surprise was complete. Hundreds of dark figures sprang from the woods and closed in on Lower Toulgas.

Had the Bolsheviks been Germans, they would have immediately rushed the Canadian guns, and the story of Toulgas would have been one of massacre. They did rush the guns, but not until it was too late. The march through the forest had been an exhausting one, and the Bolshevik soldiers were very tired and very hungry. A few critical moments were spent searching the houses of the captured village. One of the Commanders, Melochofski, a stalwart giant of a man, with a high, black fur hat, entered our hospital billet, and flourishing his arms, gave a loud-voiced order to kill the invalided soldiers. The British medical N.C.O., with rare tact and extraordinary presence of mind, placed rations and two jugs of rum before the big Bolshevik leader, who helped himself liberally to the spirits and under their benign influence momentarily forgot about the execution.

Probably in this way and in ransacking Lower Toulgas, not over three minutes were lost, but never were three minutes more costly, for during that time the Canadians swung round their guns, and, when the Russians rallied to renew the attack, they were met by muzzle bursts.

Nearly a hundred years before, at Wilma,[35] the iron veterans of the Grand Army had been shaken by that blood chilling Hourra! Hourra! of charging Russians; but now it only made those leather faced men at the guns laugh with the wild, delirious delight that comes only to the born fighting man, then only when the fight is at its height. They swore fine, full chested, Canadian blasphemies that were a glory to hear, crammed shrapnel into their guns, and turned terrible blasts into the incoming masses that exploded among them and shattered them into ghastly dismembered corpses and hurled blood and human flesh wide in the air in sickening, splattering atoms. While all the time the American sergeant and his single squad kept up an incessant fire with his Lewis automatic, and those Canadians who were not hit, and were not needed at the guns,

35   The Battle of Vilnius in December 1812. The Cossacks drove the French out of the city and west into Poland.

worked the bolts of their rifles with the energy of fiends, so that the crackling of small arms sounded like the bursts of machine gun fire from the emplacements, and deceived the Bolsheviks, who thought it was the fire of machine guns. These Canadians had used the rifle often in the untracked places of the Western World, were well schooled in marksmanship, and now when the target loomed big and at extremely short range, they covered the ground with dead.

The mere weight of those approaching great numbers would have shaken and turned ordinary troops, for the onslaught was not stopped until less than fifty yards from the guns; but the Canadians were not ordinary men and they gave not the slightest hope of being turned. They would have stood by with their bayonets to the last, and when the Bolsheviks saw the unyielding determination of these Western savages, to whom fear seemed unborn, and knew that more devastating death storms of shrapnel awaited further advance, their morale broke down, the front wave hesitated, panic spread with telepathic swiftness, and in the control of overpowering fear, the whole force bolted and scampered like rabbits to the covering trees. There they were rounded together by the remaining commissars, and from places of concealment directed a hot fire on the guns.

So quickly were they reorganized that fifteen minutes after the assault had been turned back, the Company of Royal Scots, hurrying across an open field to the support, were subjected to such a blighting fire that the ground was strewn with the huddled figures of their dead and wounded.

As the day advanced the chief commander of the Bolsheviks was killed and three other commissars were picked off and killed. The march through the marshy forests had been made at tremendous toll in vitality, the advantage of surprise had now passed, rations were running low, and, unless the attack could be pressed with renewed forces, there would be another bivouac in the wet and cold, for the Canadian devils watched Lower Toulgas, and, at the first sign of occupancy, hammered and pounded and shook the houses with high explosive until they were untenable utterly. During the afternoon an American force from the center village pushed back a band of riflemen that hung at the fringe of the woods, and, as evening fell, the enemy fire grew less sustained and it was evident that unless reinforcements arrived, the attack would fail. But hours passed and no reinforcements. The rifle reports sounded more and more erratic, and, as the night wore on, there was only the sporadic crack of a few snipers in the rear woods, who held on hopefully waiting for the supports that never came.

Prisoners said there were six hundred and fifty in this rear attack and an equal number had taken the upper village, where they kept up a steady volley fire, but seemed to wait upon success of the rear party before storming our fortifications. Therefore, far forward in the blackness of the night, the Canadians sent forth two salvos, to let this frontal attacking force know that the guns were intact and that a fight was waiting beside them.

So ended the first day of the battle of Armistice Day. There was firing all through the night from Upper Toulgas, and luminous flares burst startlingly from unexpected places in the blackness, but after the failure of the rear movement, no further sustained and determined attack was attempted.

When a patrol from the garrison entered Lower Toulgas the next morning, men nerved themselves for a fearful grewsome spectacle in the hospital billet; but lo, their comrades were unharmed, and a woman in the uniform of a Bolshevik soldier was caring for them as well as the enemy wounded. She had come with her sweetheart, Melochofski, the thirty miles from Seltzo—Lady Olga,[36] as the soldiers called her—and had bivouacked the two cold nights with the soldiers in the woods and swamps. She saved the lives of our injured men by pleading with Melochofski. Later she ministered to him as he died in the same hospital room where he would have witnessed his helpless enemies die.

She was a member of the Battalion of Death, this extraordinary woman, of intelligent, almost beautiful appearance. Madame Botchkoreva also had been a member of the Battalion of Death,[37] so named because it chose to die rather than betray Holy Russia. Madame Botchkoreva, who had come with the American soldiers on the transports from America, and had spoken to them on shipboard so eloquently and so movingly of her country and its sacred,

36   The nickname "Lady Olga" for the nurse refers to Grand Duchess Olga, eldest daughter of the Tsar, who served as a nurse in military hospitals. She was murdered along with the other Romanovs. She was canonized as a saint of the Orthodox Church.
37   Maria Bochkareva (1889-1920) when the war started, she tried to enlist. She was rejected, but sent a telegram to the Tsar asking for personal permission. He granted this. She was deployed and saved over dozens of soldiers. After the February Revolution in 1917, she suggested creating an all-female unit and Kerensky approved. This was the 'battalion of death.'

unshakable loyalty to the Allied cause, was said to have interceded with President Wilson, urged the sending of American troops to succor afflicted Russia, and prevailed upon the President.

American soldiers had already witnessed grotesque inconsistencies in this strange campaign. After the first fight they picked up shell fragments with the letters "U.S.A.," and learned that all, or nearly all, the Bolshevik ammunition was manufactured in their own country. They were told that they had been commissioned to safeguard valuable war supplies, and, coming to Archangel, had seen the great warehouses there destitute of those supplies. Now they were mystified by Lady Olga, who fought against Madame Botchkoreva in this baffling Russian war. Who was the greater patriot? Each a soldier in the uniform of her country, each had plighted her heart to beloved Russia, each had taken solemn oath to defend her country until death; and both now thought they were offering their lives for the defense of that country!

In this rear attack, one hundred Soviets were killed, many more wounded, many taken prisoners, a few rejoined their comrades at Upper Toulgas, and the rest faded in the forest and were lost. Weeks afterwards, the villagers at Nitzni Kitsa, fifty miles to the west, told of three Bolshevik soldiers who came to their village in a crazed condition, clad in rags, and half starved, babbling an incoherent story of the frightful battle of Toulgas on Armistice Day, and of hundreds of their comrades, lost in the woods and perishing in the treacherous quagmire of the swamps.

Following Armistice Day, early the next morning there was a flash at the bend of the river beyond Upper Toulgas, then the screaming passage of a shell, and the dull, vibrating, smashing roar of high explosive as it struck near the bridge. Two enemy gunboats were seen mounted with three inch and six inch guns. Further up the river and beyond sight was still another craft with six inch guns. Concealed among the trees, just on the edge of the clearing before Upper Toulgas, was a complete Bolshevik Field Battery, and these combined cannon now concentrated on the blockhouse that guarded the bridge. Shells, tossing geysers of dirt and debris, struck all around, and ploughed a deep circular furrow within a radius of five yards of the death house, where seven Americans sat with blanched faces and set teeth, counting the seconds between the hideous successive whine of the plunging shells, and waiting silently for certain destruction. At the edge of Upper Toulgas, Bolshevik infantry stood crouched for the dash, watching for the strongpoint to collapse under the terrific pommeling bombardment.

A stack of hay was near the important post, where a shell smashed, scattered the hay to right and left, and clogged the loophole that outlooked to the enemy position. The American sergeant in command sprang from the blockhouse, snatched the obscuring hay, and was back again, while bullets from the amazed Bolsheviks spurted inches over his head.

Again the same thing happened, and again the sergeant, Floyd A. Wallace, with as noble an exhibition of cool, deliberate courage as man is capable, went out to clear the covered loophole, and did clear it, but he crawled back with a hole in his tunic from a machine gun, and his drab coat was soaked deep red from a grievous wound.

It was noon when the blockhouse was hit. It crumpled like paper under the impact, and one man, drenched with a welter of blood, was seen to drag himself from the wreckage and crawl back to the priest's house. I saw this man on the deck of the transport when the Americans were leaving Archangel in June, every soldier radiant at the prospect of farewell to the army and Russia, and going home, but he had not yet learned to smile, and written on his face and deep in his eyes was the look of one who has gazed at hell.

When the bridge post was knocked out, one American, carrying a reserved Lewis gun, followed by two more each with panniers of ammunition, rushed from the house back of the church, and the three, dashing a few yards at a time, then throwing themselves flat on their faces, made the cover of a trench by the side of the priest's house, and, when the Bolsheviks came forward to the bridge, scattered them with a heavy fire.

In the emergency, a Vickers gun was hastily barricaded against a church window that looked down on the bridge. A platoon had cone down the hill from the center village when it was seen that the blockhouse could not survive, and, using the skirmish tactics of the Indian, had passed through a tempest of rifle and machine gun bullets to the billet house, and reached the church. These were only a few instances of brilliant initiative. Nowhere than at Toulgas during the battle of Armistice Day was there better truth of that French saying during the war: "Every American private soldier is an officer."

Several times the Bolsheviks felt out the bridge, and the commissars in rear could be heard urging their men to the attack, but each time they drew back before the heavy, well directed fire of the Americans, and, although the artillery smashed the white church and made of the priest's house a rent and tattered ruin, the defense held at every point till with merciful darkness the gunboats ceased their cursed belching, the guns in the forward woods subsided to blessed silence, and, screened by the shielding night, the Americans were able to bring in their wounded and send relief to those who had stood at the most exposed posts without rations or water for many long hours.

On the third day of battle, the Bolshevik batteries were augmented by two six inch guns brought down river from Seltzo to Andreevskaya, and all guns as throughout the first two days stayed safely beyond the furthermost range of our feeble three inch pieces. Despairing of breaking down the obstinate defense of the bridge, the bombardment shifted to our fortifications on the forest flank of the center village, and here for hours high explosive projectiles and clouds of shrapnel fell at the rate of one shell every fifteen seconds, ranging from the strongpoints that guarded attack from the direction of the woods, to a row of huts on the side hill close by, where a platoon was quartered as a reserve for these outposts.

Hardly had the Americans withdrawn from one of these huts, when its roof was smashed with deafening explosion, and then bolts struck right and left with stunning rapidity like raging messages from hell, flinging debris and dirt and fragments of wood in wild disorder that fell down upon the prostrate men crouching in a nearby fold of ground. The houses on the hill were raked through and through and many became a chaos of splintered timbers; the air was stabbed by the sibilant, vindictive snarl of the shells, fluttered and throbbed with their violent passage, the ground trembled in quaking travail; shrapnel burst in gray clouds, fell rattling on the house roofs or plumped down to the wet ground with suggestive vicious thuds, and the cumulative effect of successive thunderclap detonations was like a physical pommeling on the brain.

But through it all the Americans held fast, clinging to sanity by sheer point of a desperate wilfulness and facing the Bolshevik infantry men with unwavering front, so that they dared not show themselves and were still back in the forest when night came to heal the hideous turmoil of the day and still the shaking salvos that stormed through every hour of light, and would be renewed at first dawn, for the Bolsheviks never relented in their determination to take the village Toulgas.

The great Trotsky himself directed the attack. Prisoners said that, stationed like Napoleon on one of the river craft, he watched the battle from afar. The Soviet leader made an address to his soldiers and told them that he intended to keep hammering at Toulgas if it took all winter to break down resistance of the garrison. The battle was fought on the first birthday anniversary of the Bolshevik revolution, and its objective was to sweep through the Allies' lines to Bereznik, where the soldiers were promised many gifts from the valuable stores there.

On the evening of this third day we took an appraisal of our fast failing resources and estimated the prospect of a further stand. If the attack had settled to a siege, it looked as if there was small hope ahead, for a quarter of the little company had been hit, and those who remained were hollow-eyed from fatigue, so weary that they staggered like drunken men. All night long, enemy patrols prowled about the defenses, sounding them for a weak point, rifles cracked and snapped and through the black sleepless hours, machine guns beat the devil's own tattoo.

There was a tacit understanding in the way each man eyed his mate that when the fortifications fell there would be a street fight in the center village and the Bolsheviks would take no prisoners. These men from Michigan and Wisconsin had come from Camp Custer, and, when the trial came, Custer's spirit would triumph over flesh and live again the glory of the Little Big Horn. Likewise in those fighting ranks were heirs of Cromwell's men and a host of sires whose imperishable battle deeds have risen to the heights of gods the strength of mother England's fighting men. So there was no thought of surrendering Toulgas, and evacuation was entirely out of the question. If the Bolsheviks were bent upon a determined siege, they could bring fresh levies of men and new guns from their Dvina Headquarters at Krasnoborsk, a short distance from Seltzo; but Toulgas had no new guns to draw upon, and there were no supports and no reserves for Toulgas.

Our Command decided that the only hope lay in a bold counterstroke. The Scots relieved the Americans at the outposts, and in the murk of early morning, on the fourth day of battle, the American company crept through the noiseless forest and surrounded an observation post in the woods on the flank off Upper Toulgas. Several Bolsheviks were killed and the rest fled to the enemy village in panic, with the report of a great force which had overwhelmed them. The observation post with many rounds of small arms ammunition was set afire, the explosions sounded like the musketry of a regiment, and the tired and discouraged Bolsheviks thought it was a fresh regiment firing unseen from the unknown depths of the forest.

Fortune plays a great part in war, and uncertainty accounts for many things that appear inexplicable reviewed from the comfortable distance of peace; perhaps the most important information that can come to a commanding officer is knowledge of enemy strength and his fighting morale, and the Bolsheviks had no such information. They had lost their Chief Commander Foukes in this forest counter-attack, and a message from him, found on the body of a runner who was trying to reach Upper Toulgas, read:

We are in the lowest village. One steamer coming up river— perhaps reinforcements. Attack more vigorously. Melochofski and Murafski are killed. If you do not attack I cannot hold on, and retreat is impossible. 11th November, 1918. 12:30 P.M.

FOUKES.

With Foukes, four of the five commissars had been killed, and now when the frightened survivors of the detached outpost spread the alarm of overwhelming numbers of Americanskis in the forest, the Bolsheviks were seen fleeing Upper Toulgas in skeltering disorder.

The Americans dared not pursue, for to do so would have revealed their true strength, and they were outnumbered four to one. Besides, they were too elated at being rid of the enemy to give him the chance to return to the attack. They contented themselves with taking prisoner those stragglers who could not keep pace with the leaderless rabble that dispersed into the forest.

A row of houses isolated near the stream at the edge of Upper Toulgas was suspected of being the dwelling place of unfriendly peasants. The Bolsheviks used these houses as vantage points for sharpshooters, and in the counter combat a number of prisoners were taken from them, so now, when we gained the upper hand, "sniper's row" of huts was condemned, the peasants were cast out with their scanty possessions, and as the first snow filled the air and spread an apron over the drab colored ground, the homes of their fathers became a sea of crackling flames, and the poor moujiks, women and children sobbing hysterically, and men with mute sadness and uncomprehending resignation on their bearded faces, set forth to begin life anew.

The prisoners taken in this battle of Armistice Day, all except one, expressed no martyr's devotion to the cause of the Soviets. Some spoke of being impressed in the Red army at the point of the bayonet, and being kept in the ranks by the same argument. Others said that they had joined to escape starvation, and there appeared to be something plausible in this assertion for as far as we had gone into the interior the people of the Archangel villages were in desperate want. The Bolsheviks had commandeered all available food supplies which at best were not bountiful, barely sufficient to sustain the life of the villages through the long cold winter; a few potatoes with a little wheat which the peasants had cached in forest dugouts sustained life in some manner. Later had not the Allies doled out rations of flour and other food stuffs from Archangel, many in the Province would have perished of slow starvation during that winter of 1919.

The ration of the Bolshevik army was ample enough; a portion that looked princely to the moujik: a funt (fourteen ounces) of meat, one and three-quarters funts of bread, with tea, sugar and tobacco for every soldier.

If the stories of the prisoners were true and not inspired by motives of gaining sympathy, one could believe those Russians of the intelligencia who asserted that the Bolshevik party was a minority party of terrorism, and that very few Russians were ardent Soviets.

Even Lenin himself, once said that of every one hundred Bolsheviks fifty were knaves, forty fools, and probably only one a sincere follower.

Two highly cultivated artillery officers, who had held commissions in the Imperial Army, gave themselves up shortly after the battle of Armistice Day and told a tale of being forced into the Bolshevik army by the threat to kill their families if they refused. They said that all Bolshevik officers were ceaselessly observed by spies who were quick to report to Staff Headquarters the slightest symptom of a wayward disposition, or the suspicion of any gesture of mutiny.

Few of the prisoners wore any regulation military uniforms. In appearance there was nothing, except the carrying of firearms, to distinguish them from the moujiks of the villages. Both were clad in like valenkas, or felt boots, dirty, gray, curled, high fur hats, shapeless dun-colored tunics. Many of the villagers were in sympathy with the Soviets, and despite all vigilance, there was an active system of espionage between many moujiks and the Bolshevik leaders with which it was impossible to cope. Our intelligence received information that the rear attacking party had been conducted to our lines by a prominent resident of Toulgas, and sometimes the enemy showed amazing knowledge of our forces and the state of our fortifications that must have come from those in whose houses we dwelt as unwelcome guests.

There was but brief respite after the four days' battle of Armistice Day, for the American engineers set all hands vigorously to work on the winter defenses. Around the center village, blockhouses were built on the forest flank, and at front and rear at points distanced from two to three hundred yards one from the other. Coils of barbed wire were transported over the snow from Bereznik and strung in wire aprons between the strong points. Every blockhouse had an automatic rifle or a machine gun, and some at the more important posts had two, all targeted and trained to lay down a devastating, enfilade fire along the connecting wire barriers. A few Colt machine guns that were air cooled arrived, and helped the morale immensely, for they had no difficulty functioning in the very low temperatures. Then, when there was more time, the blockhouses were reconstructed with heavy timbers and piled high with sand so that they became bomb proof to anything except the explosion of a six inch shell, and even along the unfeared river bank there were placed two small blockhouses with machine guns.

When the snow mounted high and icy winds stung with the sting of wasps, Toulgas had become a fortress, well nigh impregnable, unless her defenses were penetrated from within, or the attack came in hopelessly overpowering numbers.

But scarce had all this preparation commenced, when came glorious news of the Armistice. The war was ended, and it was taken as a matter of course that the coming peace would extend to the war of the Arctic Circle.

From the outset the soldiers never had any rampant enthusiasm in this strange conflict with its motives of mystery, but while

the struggle in France went on they stilled their questioning
doubts and followed the work set out for them by their officers
in the uncertain belief that somewhere back of the scenes at Paris
or London or Washington those in the high places had charted a
wise policy beyond the comprehension of a common soldier; and
that in same devious, undisclosed way the campaign in Russia was
necessary, was playing its inexplicable part in completing the defeat
of the Germans. Even when weeks elapsed and no announcement
of change in policy was forthcoming, the men were patient and did
not complain. But when at the end of November, Consul General
Poole sent word from Archangel that the Americans in North Russia would continue at their tasks to the end, knowledge came to the
soldier with stunning reality that the great struggle in which he was
prepared to die had no relation to the war with Russia, in which
he probably would die, that he was engaged in a war which had no
assignable reason for its being, in which many of his companions
had already been killed, and the end was not in sight.

The uncertainty, the isolation of the distant snowbound fronts,
the ever present prospect of being trapped by enemy occupation of
the villages along the extended communication line, and now that
the excitement of the fight had waned, the depressing monotony
of the days ground down the spirit of the men. They commenced
to lose heart. Life became a very stale, flat, drab thing in the vast
stretches of cheerless snow reaching far across the river to the
murky, brooding skies and the encompassing sheeted forests, so
ghostly and so still, where death prowled in the shadows and the
sinking realization came home of no supports or reserves along the
two hundred miles of winding winter road to Archangel.

Week follows week, and November goes by, and December, and
no word comes from the War Department. No reassuring message
to the perplexed Commander-in-Chief, defining the purposes of
the war, its duration, when relief will come. No word comes and the
soldier is left to think that he has been abandoned by his country
and left to rot on the barren snow wastes of Arctic Russia.

Men move about wintered Toulgas emitting great clouds of
vapored breath, shuffling over the snow in the clumsy Shakleton
Arctic boots, wrapped in great coats against the bitter, deadly cold;
on their faces the condemned look of felons from whom all hope
has fled.

In the dismal huts of the village soldiers are packed with the

crowded moujik families like herded animals, where the atmosphere is dank and pestilent with an odor like stale fish. Filth is on
the floor and vermin creep from the cracks and crevices of the log
walls.

In December and January there are only a few hours of feeble
shadowing light, then tragic blackness blots out the snows and
the mournful woods and the skies of melodrama. With night the
tiny windows are shrouded with board coverings, a candle flickers
in the low ceiling room, unless the issue is exhausted, then a bully
beef can is produced, filled with bacon grease and an improvised
rag wick which flutters a hesitant glimmering through the heavy
gloom.

There through the long dark unwholesome hours, the Americans sit and think thoughts more black than the outside night. Red,
hateful, revolutionary thoughts like those of the maddened mob
that rushed Louis Seize[38] to the guillotine, and that would threaten
the stability of any nation. Black thoughts of their country and the
smug, pompous statesmen who with sonorous patriotic phrases had
sent them to exile; of the casual people at home and their damned
complacency and their outlook on war as a gorgeous heraldry of
youth, a gay, romantic adventure.

Sometimes it almost seemed as if malignant Bolshevism had
poisoned the air, for once in February when the situation looked
worst and nothing seemed certain except annihilation for the
whole garrison, the American soldiers at Toulgas threatened, unless
promised early relief, to walk out like disgruntled factory hands.
The same thing, but with a more serious aspect, occurred in an
American company at Archangel; and the French on the Railway
had, at first rumor of the Armistice, flatly deserted and returned to
Archangel. At Kodish a company of British refused to fight further in this indefinite war, and among the first conscripted Russian
troops there was serious mutiny resulting in much bloodshed.

But there was nothing mutinous in this expression of opinion
at Toulgas. It seemed the only course to civilian soldiers who were
schooled in strikes under an industrial system where the strike has
always been the concerted expression of disapproval by those who
toil in the ranks. When the nature of a mutiny was explained to
these men, they felt a burning shame for what they had done so
unwilfully, and never again, throughout the many discouraging,

38   King Louis XVI (1754-1793) the French King executed in the French
Revolution.

hopeless days that followed, was there the smallest hint of protest from these civilian American soldiers.

When the days were shortest, the commissary transport broke down, and for a time the principal ration was corned beef that was frozen in the tin, and a nauseating mixture of vegetables and stewed meat that had been alternately frozen and thawed in the tin, and when eaten, gave some loathsome skin diseases and others dysentery.

Cooking and eating were the only breaks in the melancholy monotony; there was no diversion, no relaxation, no recreation, and the divine gift of humor which was the salvation of the Western soldier, was denied to the soldier of North Russia, for humor springs from buoyant spirits, the wells of radiant health, and the Americans on the Dvina were so physically depleted that in February the medical officer of the First Battalion reported that one-third of all those on active duty should be committed to the hospital without delay. But these sickened soldiers could not be sent to the hospital without abandoning the undermanned posts that guarded the garrison.

Robbed of physical resistance and broken in spirit it was pitiful to see strong men and brave men become shrinking cowards, filled with a vague, sapping dread, under the uninterrupted strain and the depressing influence of the long nights. Fidgety sentinels were constantly seeing lurking Bolsheviks conjured by their morbid imagination from the menacing shadows of the woods, and there was an epidemic of accidental self-inflicted wounds, which always occurred at the ticklish, unsupported, advanced positions.

The doctors pronounced many as cases of neurasthenia induced by much loss of sleep, unbroken fatigue, and continual drain upon the nervous forces. They looked solemn and dubious and said it was demanding too much of human endurance to expect the defense to hold on without relief through the many winter days that stretched ahead.

One January night, terrible in the severity of its cold, all hands "stood to" and waited for the rush from the woods, for sentinels had heard the muttering of many voices and had caught the movement of bodies among the trees; / but no attack developed, and in the morning the tracks of timber wolves were found approaching almost to our wire, where the pack had stopped to sniff the scent from these strange tenanted loghouses, standing apart on the snow, like outcasts of the village.

The few sentinels kept far in advance at the front village were always having jumping nerves, and robbing exhausted men of precious sleep; but once in truth they were nearly surrounded during the night and escaped by a miracle. So it was decided to burn the houses, as "sniper's row" had been burned in November. Some two hundred peasants were turned out in the snow, and Upper Toulgas became a dirty smudge on the whitened plain over which our range of visibility extended far to the forward woods, and our field of fire was increased comfortingly.

The High Command passed out word that Arctic conditions would preclude any active fighting, but the prisoners spoke differently. They said that the Bolshevik Staff expected the Allied soldiers to die like flies in the cold winter, that the enemy intended to strike when the cold was most bitter, the snow deepest, and so they did.

In January, with a temperature forty degrees below zero Fahrenheit, at midnight, Bolshevik batteries from across the Dvina commenced shelling Toulgas, and continued for fifteen minutes a bombardment that went wild in the dark and struck harmlessly far from our works.

Directly the last shell had been fired, enemy infantry advanced in the open and rushed our front posts. In the darkness there was frantic, wild fighting and struggling in the deep snow, shrill yells and a confused babble in a foreign language, the hideous moans of the wounded, the ringing commands of the commissars in rear, urging their men forward to sure death, and the prolonged explosions of machine guns spurting a rain of bullets over the heads of the attackers to warn them of a death that waited in rear if they turned back.

In two hours the force of the assault was spent, the last shot had been fired, and the snow before one of the blockhouses, where enfilading fire had cut up the attack, was covered with Bolshevik bodies. The fight was an uneven one, for the Americans in the blockhouses fired from bullet proof cover and were sheltered from the weather; but the Bolsheviks had to advance against barbed wire, struggle in the snow against targeted machine guns and had no protection from paralyzing cold. Many of the prisoners were so badly frostbitten that arms and feet were amputated to save their lives.

In February, acting in cooperation with the enemy offensive on the Vaga, a large force of fresh troops composed mainly of the Eighty-second Tarasovo regiment, who knew nothing of the reputation of Toulgas and the fate of other attacking parties, waded through the cold snow forests, clad in white smocks to blend with the color of the ground, floundered up to our lines in the impenetrable night, and were not discovered until they were engaged in cutting the wire between two blockhouses. They were fairly trapped then between the enfilading fire of two sets of machine guns and suffered fearful carnage before they fought their bloody way back wading ponderously through the deep snow to the forest.

Some of the dead came abruptly to life and gave themselves up when a search was made of the bodies next morning; horribly frozen by exposure, they said they preferred an uncertain chance of life at the hands of the Englishskis and Americanskis, to the certain chance of death in a further attempt to conquer Toulgas.

After this sanguinary fight, the Bolshevik soldiers met in a great assemblage, made bitter speeches against the Commander who had led them to disaster, and resolutions were passed which threatened death to any commissar who insisted on another assault of Toulgas and the fighting fiends who defended it.

So this village, far up the Dvina, was no longer the prey for wild midnight sorties and desperate melodramatic clashes in the deep snow, and there might have been comparative peace for the garrison were it not for adherence to those cardinal precepts of military orthodoxy that aggressive contact with the enemy must be always maintained and reconnaissance is vital to a successful combat campaign. It was to conform to these inflexible precepts of the military that patrols left Toulgas seeking for Bolsheviks. Sometimes they went forth on webfooted snowshoes, and scouted the forest far on the threatening flank to discover whether the enemy had found some new method to approach our positions, and then they served a useful purpose. But the customary patrol party was the one that went out every day, a band of three or four, along a trail of padded snow just wide enough for a single file, that led through the front forest, five miles to the nearest enemy position at Zastrovia.

A hunter can understand this tracked snow trail. It was like a game runway that leads to a salt lick, fresh signs show that deer pass every day, and it is only a question of time until the hunter gets his chance for the fatal shot.

Sometimes, by the mere coincidence of fate, a patrol would turn about in the trail and start back towards friendly lines, when a machine gun would snap and crack and a rush of bullets sing harmlessly high, where another hundred yards meant death from the ambuscade; and often the scouts would come to the hidden waiting spot where imprints in the snow left the story of a large Bolshevik force that had stayed long, but, overcome by the cold, had been forced to quit the death hunt.

Often the Bolsheviks would leave bundles of propaganda on these patrol paths, much of it written in English, inciting British and American soldiers to mutiny, to kill their officers and join the Soviets in a revolution for the world wide supremacy of the proletariat Death walked these white runways. Death, and his romantic partner, Chance. But the color of youth had vanished before dour, wan reality with the soldier of North Russia, and the romance of Chance was lost on him. Yet it was strange how often men could walk these suicidal paths and escape unscathed. The goddess was kind, she visited them with benevolent mood, save a few times such as once in March, when from a party of seven, only one got back to tell of the fatal ambush.

When a platoon hurries out to pick up some sign of the others, it is caught in the open at Upper Toulgas, pocketed from the supporting fire of our own lines. There in the open snow, and denied all cover, the men are trapped like condemned animals. They flatten on the snow and fire at an unseen foe that pelts a withering fire from behind trees three hundred yards on a quartering forward flank; bullets whip the snow beside them and sweep by in such a storm that the air whimpers and cries aloud like a tortured living thing. At the end of three hours snow clogging in rifle breeches has frozen solid and they can shoot no more. Then, when it looks as if all were lost, the last man on the line gets back to the artillery, but is so winded and funked by his experience that his directions are a confused babble and the artillery opens up at risk of hitting our own men, shrapnel bursts in front of the platoon, the murdering fire from the clump of trees slackens, and the officer is able to withdraw his men to a God-given dip in the ground, all that are left of them, for out on the white snow still stretches a crumpled drab colored line; some lie very still, others writhe in the agony of grievous or fatal wounds.

Two days after this shambles of the snows, an officer and three men were met, on the forest runway to Zastrovia, by the fire of a large force of Bolsheviks, but until the day the Americans left Toulgas, there was no abatement of the perilous policy of patrols in this undefined war, where the loss of every life seemed sacrilegious sacrifice.

And this amazing campaign so prodigal of men's lives continued through the lengthening winter days.

At the end of March the sun had mounted high, and the snows were fields of myriad dazzling diamonds. A new fresh fragrance filled the air, and brought the promise of vague, perceptible hope. Spring was coming with the sun, and the renewal of youth would not be denied.

Then the Headquarters of the American Expeditionary Force took cognizance of the war with Russia and sent a general officer to command the forces from Archangel.

Then the Secretary of War announced that no more troops would be sent, and the units there withdrawn.

This was the end, but the Americans did not know it. The Royal Scots came to take over the defenses, the old Category Bs, with their wound stripes, their traditional, cockney jauntiness and just a hint of superiority in their eyes for the Yanks who were leaving the show.

It was strange how that night the winter's harshness relented in the gentle lulling wind, and in the luminous spell of the limpid moon, weary, war-worn Toulgas was at peace, sleeping, in unbroken white stillness.

Far up the sloping hill the rude silhouette of the center village is etched against a starlighted sky. Forward the church, shell gashed and mutilated, with its grotesque minarets, and the moon, a pendulous globe of living fire. Clear in the lucid light is the hard contested bridge, that means so little and yet so much; beyond, the charred ruins of the sacrificed village, and, still farther, the somber, gloomy forest. Vividly white gleams the church beneath the steely mystic moon, but whiter than the church or moon are the endless wastes of immaculate, unmarred snows that reach across the great river to the lurking darkness of the distant shore and abroad to the sinister shadows of crested trees.

This is Russia of the American soldier—a cluster of dirty huts, dominated by the severe white church, and, encircling all, fields and fields of spotless snows; Russia, terrible in the grasp of devastating Arctic cold; the squalor and fulsome filth of the villages; the moujik, his mild eyes, his patient bearded face—the gray drudgery and gaping ignorance of his starved life; the little shaggy pony, docile and uncomplaining in winds, icy as the breath of the sepulcher; Russia, her dread mystery, and that intangible quality of melodrama that throngs the air, and lingers in the air, persistently haunts the spirit, and is as consciously perceptible as the dirty villages, the white church, and the grief-laden skies.

It was not until nearly June the Americans were told that their bizarre service to their country was at an end. They were to go by slow stages back through the Dvina villages, always within call in case of dire need. But at last the purple day comes, and they are going home. A troop ship off among the ice floes of the White Sea toils westward, and upon its decks is a throng of soldiers who gaze with equivocal valediction upon the failing Russian coast, which mingles imperceptibly with the distant haze, and so passes like this shameful war to the bourne of memory's empire. The fairy rumor has come true, the Americans are going home.

34607   Pvt.George Dhimian, company G, 339th Inf. and Pvt.Marvin Mock,
         Co.F,339th Inf.,carrying water in the native fashion from the
         Dwina river to barracks. It is boiled before using. Archangel,Russia.
         Oct.7,1918,

The only means of transportation after the rivers were closed

THE VAGA

27th Oct., 1918

Dear Colonel Stewart:

I understand you have very little information of the situation up here. I have very little myself, and what I get is usually from rumors unless I go to British Hdqrs and ask for it which I do not care to do.

.... The commander of Force C has my Bn scattered so much there is only one company in a place. Have two companies under my orders Co A is up the river about 25 versts from here Co C is at this place and one Plt of Co A. Co B is over on the Dvina and Co D is with Force D about half way to Archangel between the river and the railroad.

.... Suppose part of us will winter here, but do not know yet. . . .

Excerpts from letter written from Shenkurst on the Vaga, by Lieutenant Colonel John B. Corbley to Colonel George E. Stewart, Commanding Officer, 339th United States Infantry, Archangel, Russia.

"In North Russia, Shenkurst has been abandoned and the Allies are in a precarious position. The country is apt to hear much of these American battalions of North Russia, whether they live or die. If they live, it will be only after an heroic struggle with two fierce enemies—man and nature. If they die, it will only be after they have expended the last ounce of strength and the last cartridge."

The Washington Post, 28th January, 1919.

"Shenkurst has been evacuated and we are greatly outnumbered, but there is not the slightest reason for anxiety. New positions have been occupied a little further north. The Archangel expedition is quite safe, and always has been safe."

The London Times, 28th January, 1919.

THE VAGA

THE meagre numbers of the Railway had been irreparably spent by the establishment of the Onega force, on the west, and a like outguard at Seletskoe on the east, with its right and left wings, Kodish and Shred Mekrenga.

Now, as it followed up the Dvina, in the same manner, the dubious, striking power of the River Column was lost by the output along the tenuous, weaving waterway of many communicating posts, that like great drops of heart blood from a mortal hurt, wasted its vitality and drained its strength, until it could go no further.

These posts, like Indian blockhouses of frontier days, were strung along the river course nearly to far Archangel, and in them, insignificant detachments, with the grim, quiet resolution of the frontier men, and the steady, reliant nerve of the frontier men, safeguarded the backward way, where always silent, winter darkness held ceaseless, dire, ominous threats.

In the Shred Mekrenga offensive of January, when the enemy sought to cut off the River Column from its base, he launched a venomous attack at one of these river posts far back at Morjagorskaya, but the British garrison held without flinching and saved the communications by a narrow margin.

By this process of dispatching numerous, guarding detachments throughout the province, the Allied forces, utterly trivial at the outset, became so dispersed that the "offensive war" swiftly degenerated into a disjunctive, raiding excursion, and the invasion, instead of striking the Red Bolos with terror and chasing them like scurrying quail to cover, was regarded by the enemy with contempt, even derision. The Bolshevik soldiers, at first panicky, soon overcame their fear, and when their leaders saw that no reinforcements could come through the frozen north port, they assumed an attitude of aggressive defiance, and were ever conducting raids, ever menacing the long, basal lines, the flanks and rear of the far separated, uncoordinated, unsupported Allied fronts. On the Dvina, hardly had the detached American company taken over the defense of the costly stores at Bereznik, when friendly natives from Shenkurst directed the observation of our Command to the danger of a rear flanking movement from that quarter, so half of the garrison was detailed up the Vaga to take possession of this city of Shenkurst in the name of "friendly intervention."

It must be said that for the most part the city welcomed, with a genuine, welcoming spirit, the coming of the foreign liberators, for many people had fled north to Shenkurst from the violent Reds at Moscow and Petrograd, who hated the intelligencia and everything else that was unproletarian, with a destructive, vehement hatred.

These people were the Russians of literature, cultivated and mannerly in appearance, soft spoken in approach, and accustomed to the niceties, the softer things of life. They wore shoes and stockings, and with a revealing hint of gawkiness, most of the rest of our unimaginative, Western habit; also they had a few of the simple delicacies on their tables that seemed like fairy gifts to the homesick, American soldiers.

The Vaga is noticeably smaller than the Dvina, and seldom exceeds a breadth of a half mile, more often it is five hundred yards, even less, and the soil through which it plows a tumid trail is soft, sandy loam, so that high, commanding bluffs have been eroded by its waters, where the villages group in almost neighborly proximity. On one of these bluff heights, stood effete Shenkurst, a generation removed from moujik poverty and enchaining ignorance, and consciously superior to the humble log huts that below north and south trailed the river. The dominating buildings, a monastery, a barracks of the Tsar, and five conspicuous churches were white as Russia's snows, and in the fall, made Shenkurst flaringly garish in its frame of tenebrious, surrounding forest.

Nearly a week of tranquillity passed with the Americans at Shenkurst, when the Staff, chafing at this prolonged unbelligerency, issued orders "to stir up the enemy," and some one hundred Americans, with fifty Allied Russian soldiers, embarked to reconnoitre the upper river.

All was uneventful, until ten miles out from Shenkurst, when suddenly an unseen fire poured from both high river embankments on the steamer bearing the unsuspecting, scouting party; there was no method of gauging the ambuscade, which judged by the volume of fire, most of which screeched harmlessly high, was far stronger than the Americans; but on the instant, the officer beached his craft on the nearest ground, the eager men scrambled over the side into the water waist deep, and engaged the enemy, who was so taken back by this unexpected action that he wilted into the forest; then, entirely undaunted, the little party moved on down the forest road, which wound south with the river, and into the sinister shadows of an unexplored, uncharted, alien country, where many signs pointed to certain, overpowering resistance, and the law of probabilities pointed to extinction.

The American in command, Captain Odjard, was more an antique Viking than a city-bred modern, and as the intrepid march continued, he never wavered in his purpose to penetrate the heart of the Bolshevik stronghold; for twenty days he kept on, despite distressing hardship, and short, iron rations, and most grievous of all, the utter absence of comforting tobacco. Reports came constantly that the enemy was intent upon the capture or destruction of the little band, Bolsheviks thronged the forward way through the forest, and every day information reached Captain Odjard that the villages in his rear were heavily garrisoned with enemy forces; most serious of all, the fast vanishing ration supplies would soon be all gone. Situations such as this search the innermost fiber of the stuff that makes for leadership. There are no precedents. A man of courage and valiant will would face about and fight his way back and perhaps die fighting. A coward would vacillate and falter in a mortal terror of indecision, and thus perish.

Stonewall Jackson and Forrest would do the genius born, unexpected thing. The Viking pressed onward, met the hostile Russians, forced them to a savage engagement, in which they lost in killed and wounded twice the number of the entire reconnoitering force, then turned about and backtracked the cleared way to the south, hastily abandoned by the Bolsheviks, in every reasonable fear of meeting the outnumbering reinforcements that surely must be coming up in support of such a bold and confident advance.

But at Ust Padenga, fifteen miles from Shenkurst, the party was stopped by a dispatch from Headquarters. It would go no farther downstream, but would act as an advanced outguard for the main Vaga position, a barricade to serve as a distant, delaying obstacle, and so render the inner post more easily defended.

For when the notion of an offensive war languished with the General Staff, and had nearly expired, it was revived a little by the theory of "an offensive defense," in which the six, widely scattered, battle fronts acted as protective tentacles, each of them in turn establishing an "offensive" outguard for Archangel, since once this virus of the " offensive defense" was inoculated in the Allied Command, it would not rest dormant, but persisted, assertive to the ultimate.

Meanwhile, Nature, flagrantly disrespectful of the military, swung the seasons in their immutable cycle. Fall made her parting courtesy, and winter with dread message and icy breath waited on the threshold.

The hope was not yet dead of the Railway Column gaining Plesetskaya, and the present objective of the Vaga force was to penetrate some eighty miles to Velsk, an important junction point of roads converging from the area of Plesetskaya, from the city of Vologda and from the Dvina.

The Railway got little further than Obozerskaya, and the little River Column, by the end of October, was at bay, fighting for life nearly two hundred miles from Kotlas, its first objective.

But before these forces had been halted, already the Vaga Expedition had gone too far, thrust out nearly one hundred miles from the Railway, and fifty miles further south than the River party, it presented inviting opportunity for enemy encirclement—a dangerous salient, projected midway between the two main Columns, and nearly three hundred miles from Archangel, by the tortuous course of the road.

The British are a bold people and it did not seem to weigh heavily with them that Shenkurst, the base of this Vaga Column, was flanked by hostile villages, where vain attempts had been made to drive out the Bolsheviks, that the city was garrisoned by locally recruited Russians, who had been tried and found wanting under fire, and whose loyalty might wane when the tide of Allied fortunes ebbed low, as soon it did.

Shenkurst must be held, and so the reconnaissance patrol, which had eluded doom only by the splendid dash of the men and brilliant leadership, stayed at Ust Padenga as an advanced outpost, and the theorists of the "offensive defense" were satisfied.

Captain Odjard took main station in a village on a precipitous cliff, that reared high from the river, and posted his Russian retainers in huts that clustered on the flat bank of the Vaga, nearly midway down the long valley that spread south to the forest.

Quartering from this second village, and much further down the valley was a third, conspicuous on another abrupt bluff, which when seen from the distance of the main post, the house tops had the picturesque appearance of toy roofs, sculptured on a pedestal.

The houses on the flat river bank stood out naked on the snow, and in case of attack, could be supported from the main position, for they were well within effective shooting range; but the other, the elevated village, was nearly a mile away, and beside it, on the west, the forest crowded perilously near; gullies were at the base of the bluff which made "dead ground" there, a series of natural trenches for an attacking party. It was a hazardous spot, the Russians would not stay in this distant, treacherous "Death's snare" on the heights; and they wagged their heads lugubriously over the few Americans who persisted in holding it. From the steep side of Headquarters' cliff, the usual wagon road descended, sent offshoots to the two south villages, and trailed off to the concealment of the lower forest.

Week succeeded week in lonely Ust Padenga, where the sad disgarnishment of this tragical, little war was seared vivid in the living consciousness of American soldiers. The Armistice came, but with it no word of enlightenment, until they were led to believe that in the general rejoicing, the stirring movement of momentous events, no heed could be given to the trifling performances of their fantastic, Arctic side show, long since forgotten in France.

Yet strange, the soldiers did not grow deeply embittered, a stoic calm came over all and they became worshippers of the Russian philosophy, *nitchevoo*, votaries of the Fates, burning frankincense at their shrine, praying favor, yet unmoved by their displeasure, indifferent to their whimsical caprice. They became atrophied men, asking nothing of the future and expecting nothing. The doctors said many were cases of neurotic disorder, and others suffered from enteritis and scabies, and ordered rest and the hospital, but the Staff waived the medical men brusquely aside and sarcastically asked who was to hold off the Bolsheviks.

During November, and shortly following the Armistice, two patrols "seeking contact," were waylaid in ambush, and from the first, only one man came back. The officer of the second might have escaped, but to do so he would have had to leave a detachment in distress, surrounded in the forest. He rather chose the hazard of death, and leading the fight, he laid down his life for his friends.

During the weeks of December and January, with their bitter cold and dismal, somber days, trees were felled about the defenses to widen the field of fire, and long, intersecting lanes were laid through the forest like swaths through a standing grain field, so that the machine guns and the automatics might hurl their spray of death at longer range, where skulked shadowed and grisly, white forms. When in the dead and quiet of the night, rockets burst from unknown quarters, flared with ghostly glare and faded in mystery behind inky, plumose silhouettes.

In the cold and the long darkness of winter, there was time for reflection for any one who would be so idle, on the defenselessness of the position, the remoteness from the base, the hordes that were massing on the road north to Shenkurst and meant soon to make "the big push."

Our Intelligence reported that in January the Sixth Bolshevik Army of the north numbered forty-five thousand seven hundred, and the dribbling replenishment of our forces that had come down the railway from open Murmansk, had far from kept pace with attrition by sickness and gunshot wounds. Disregarding our Russian Allies, we did not have six thousand men at all fronts.

By the middle of January, a blighting influence, a devastating, nether presence filled the air, like the spell of an evil spirit, and as capable of being finitely recorded as the testimony of eyes and ears. There was in the atmosphere something closely akin to that heavy, stifling calm, that in the summertime hangs over all, before the wind swoops down and the first, big, pelting raindrops fall from blackened thunder clouds, the advance guard of the drenching storm that descends to earth in howling, unrestrained fury.

All at lone Ust Padenga knew the storm was coming, it was only a question of where it would strike. On the 19th day of January, the dispositions were these: a platoon of Americans held the village on the pedestal, fifty-four allied Russians were in the village on the flat below, and the main body of Americans, some two hundred strong, two Field Pieces, one One Pounder of Russian design, one Pom Pom and forty Russian artillerymen (who funked in the first fight and were relieved by Canadians), were in the backward village on the high bluff.

At dawn, for one hour, enemy batteries from across the Vaga shelled the foremost position on the elevated ground, then suddenly ceased firing, and like grotesque Jacks in the Box, swarms of white-clad Bolsheviks arose by magic from the concealment of the ravines. A succession of long, white lines came from the close forest, and across the open snow of the Vaga came still more advancing, white-clothed men.

Against such bulked masses, resistance was impossible. Three machine guns, burst after burst, tore rending gaps in the coming lines, but they merely welded and kept on.

When the last pannier of ammunition was gone, word was given to blaze a path through to the rear—and double time! And now down the steep hillside the trapped company charged, tumbling and fighting like maddened, cornered animals, until they gained a foothold on the road which stretched out bleak and coverless eight hundred yards to the main village. Some tried to make a run of it over the bottomless, intervening snows, where they struggled piteously like hobbled animals and were killed. But in most part, they dashed in frantic relays down the open road, sprinting forward a score of yards, then flattening on the ground, and so on, rushing and sprawling flat, until the fatal course was run, while every rifle from the abandoned village on the height, and the flanking forest and across the Vaga spurted death, and machine guns rattled rasping death, and bullets lashed the air with the furious cracking of ten thousand whips, or sped fluttering through the snow, and went off whimpering into space, or felled men with sledgelike blows, until the doomed way was strewn, end to end, with the prostrate forms of the fallen ones, and a pitiful few, by some fluke of luck, had gained the shielding hill.

Not ten minutes had been taken in that terrible dash through that valley of Death's shadow, and of the forty-seven who began the journey, six reached the goal of the main village. In the fearful sub-zero temperature, all of the wounded would have perished by freezing, had not a volunteer party, braving the unspeakable, barbarous Bolos (who for some reason held their fire), gone out in the open snow and brought them to shelter. Fifteen were thus accounted for, and the rest lay somewhere beyond sight, "missing in action," that ambiguous, impersonal expression of the War Department, so fraught with mingled hope and dread, harrowing fear.

When night screened the battle scene, the Allied Russians, upon their own inspiration, evacuated the village on the flat, and the next day, the unwitting Bolsheviks began the second phase of their investment of Ust Padenga. Again the artillery, even more violently than the first day, flung hurtling blasts at the deserted village, and late at day, the infantry, grotesque, bobbing objects out on the wide snow stretches, stormed the uncontested position. It was like rifle practice to shoot down those living targets, glaringly open on the white snow, and they were downed by tattering bursts of shrapnel, downed by musketry, downed by awful devastating bursts from machine guns, that moved them row upon row, until the last man had passed to the cover of this village of costly folly, and the snow was dotted with dead and wounded, which, from the distant hill, looked grotesquely like raisins stuck in an immense rice pudding.

On the third day, the surviving village, lying bare on the unsheltered top of the cliff, was the target of a barrage that searched it house to house, until many of the moujik homes were wrecks of smashed timbers, and the trail of human wreckage was a ghastly, unsightly thing. The American doctor went to death, a victim of the shells, because he would not have his wounds bound up while a single, private soldier was not relieved, but he lives with Vaga men as long as life endures, a symbol of moral grandeur and noblest self abnegation, that will ever inspire faith in the immortal, spiritual entity of man.

It was not the Viking Captain who ordered retreat from Ust Padenga. Half of his little company was gone, but he had no thought of yielding. He would have held on until the last dog was hung, if superior directions had not come from Shenkurst. He loved a fight, this antique Norseman, loved the wild, esoteric fury of it. Three times, his men threw back the Bolsheviks, and caught in a contagion of blood lust, they craved still more, maddened by battle, they took hilarious delight in seeing "the Bolos bite the snow banks."

They did not know that pitted against them was the vanguard of an army that by every objective rule of warfare should have crushed this rash, little group to utter destruction; but if Ust Padenga did not know, all at Shenkurst were fully alert to the gravity of the situation. This was the much proclaimed Bolshevik offensive, with its object, the annihilation of the Allied North Russian Expedition; and now as the full fury of the gigantic, impending assault unfolded, the "offensive defensive" theory found vindication, for at the Ust Padenga, little more than one company had stood off a regiment of the enemy.

There seemed small hope of escape for the valiant Vaga men who remained after the fourth night of the attack, when an incendiary shell fell upon the village, sending hungry, devouring flames athwart the curtain of the Russian night, till naught was left of the moujik homes save the gray ashes of "friendly intervention"; but in the confusion of concentration, the assemblage of large numbers and numerous troop movements, the retreating company glided in darkness down the center of the frozen, white covered Vaga, through the very midst of unsuspecting, enemy hosts, and two nights later, reported at Headquarters tired and half starved, the Viking leader among the casualties with a serious wound.

In Shenkurst, the beleaguered city, in point of numbers, the Slavic Battalion, nearly twelve hundred strong, was the mainstay of the garrison, but on trial in a previous attack for one of the two flanking villages, it had made a sorry showing, and in a last stand, was estimated as of uncertain, staying quality. Besides these Russians, there was one full company of American Infantry, the exhausted half company from Ust Padenga, one section of the Thirty-Eighth Canadian Field Artillery, four Two Point Nine mountain pieces, and three trench mortars.

The Bolsheviks had surrounded Shenkurst in an immense, unnumbered multitude. They had mounted one nine inch gun, two six inch guns, four Four Point Sevens and a Battery of Field Artillery, and from three-quarters of the forest commenced to batter down the buildings.

It could be only a brief time before the city would be in ruins, but even more serious was the question of provisions. They were already limited, and in case of siege, no new supply could be brought up until the breaking of the river in May.

The Bolsheviks, confident that the garrison would try to escape from Shenkurst, waited in great masses on the main north road, eager for the coming slaughter; but a native had informed the Allied Command of a secret path through the deep, snow covered swamps, and at midnight, along this unknown route, evacuation was silently effected.

Before the retreat, the Allied Russians were sent as a protective screen along a flanking trail, but scarce had the retiring movement begun, when what remained of them came rushing back in frantic haste, that was altogether unsoldierly, gasping an excited, incoherent story of how two entire companies had deserted to the enemy lines and the rest had fled in desperate fear for their lives.

Many civilians joined this bizarre, midnight march through the snow forest and swamps, and made the retreat a spectacle of wantonous disorder, as stoical men and wailing women strove heavily on, bent under the torturing weight of bundled treasures, which, under duress of fatigue, one by one were reluctantly abandoned, leaving a pathetic havoc of cluttering waste in the trail; and soldiers, weakened by much fasting and sleepless battle nights, lurched in the darkness, fell and lay in the cold snow, and had to be struck and urged on by violent means, so grateful was any surcease from further excruciating effort.

Late the next day, a merciful halt for the night was made at Shaguvari, Where a rear detached outpost of Shenkurst had been maintained, and which outnumbering, advance enemy patrols had vainly striven to dislodge. But the disheartening march was resumed in the morning, when the Bolsheviks were reported collecting in force to cut off retreat downstream. So Shaguvari was added to the sum of Russian villages fed to fires of the Allied cause and became another charred ruin on the Vaga.

At villages outside of Kitsa, twenty miles further, trenches were dug in the snow, and barricades improvised of trees, in order that the driven troops might catch their breath. And on the Dvina, now only a few miles away, new positions were taken, where the imperiled River Column could be drawn back, and the consolidated Allied forces stand embattled in a desperate last defense of Bereznik, for if Bereznik fell, all knew it meant the beginning of the siege of unfortified Archangel.

But the delaying action was prolonged beyond the most sanguine dream of hope, and at Vistafka and Yeveevskaya, Maximofskaya and Ignatevskaya, the neighboring villages of Kitsa, the Americans held out, relieved in turns by British troops, and the remaining Slavic allies, who atoned for much by a heaven bestowed blunder that saved a surrounded post of the Americans.

These places, with their unpronounceable Slavic names, will be remembered always by the Vaga men, for here during Arctic February and March days, they fought savage, bloody fights in the mounting snowdrifts, and performed deeds of sublime sacrifice and courage, that will never be known save by those who were there.

They were still at Kitsa, and had not given ground, when the first redolence of spring softened the rasping, winter winds, and made the Bolshevik Commander draw back his artillery in fear of being mired in the yielding snow roads.

Not one of the Vaga men, in the innermost counsel of his heart, had ever expected to live through that winter onslaught, and when all with quiet courage stood ready for the end, lo, the enemy abandoned the field where victory awaited, and left the battle when it had been won. This petty, strange and inexplicable war was freighted deep with countless things of mystery, but none so beyond understanding as the failure of the Bolshevik Command to follow up the capture of Shenkurst.

The feeble Allied remanent on the Vaga was reeling from the stunning blows of the massed attack, and thought of resistance all hung on the hope of saving Archangel and the life of the Expedition; but when all tensed themselves for the crucial shock, it did not come, the Bolshevik advance weakened and faltered and held back, so that the defenders, panting in terrible exhaustion, were able to suck in the air of reviving strength and hold on. When later the attacks of February and March came, they were sporadic, and lacked the fury, the sustained and vehement driving power of the first assault. Now in spring, it was too late, for Nature with sun and gentle breath had definitely won the battle for the Vaga men, and they crossed the river to safety, leaving in the black, despairing night, two villages flaming, a recessional of ill-will and destruction.

The first boast of "one Allied soldier against twenty Bolsheviks" had been made good, and the Expedition was saved, but by a precariously close margin. In no respect did the Allied Command so underestimate the enemy as in his power of military organization. The miserable "Bolo brigands" that were to have disbanded with the first punishment of Arctic cold, had raised an enormous army, which now, in late winter, exceeded one million soldiers, and the regiments that took Shenkurst must have laughed contemptuously at the undisciplined, untrained troops of the early days of the campaign.

Perhaps it will never be known why the Allies were not destroyed by these Vaga attacks. There were many villages capable of housing great numbers of soldiers south of Shenkurst, and probably in the January thrust, seven thousand five hundred to eight thousand hostile troops were quartered in them, a force that should have swept the Vaga Column before it like chaff in the storming wind, but it did not do so, and one may conjecture that the reason was because Trotsky did not care to hazard the risk of stirring the American people and the British people to an avenging and genuine war by the annihilation of the lone Allied battalions. Greater wars have been brought about by more trivial causes; but the stronger probability is that the Bolshevik soldiers revolted at the staggering slaughter of the attacks over the deep snows.

"Our losses are terrible," said one of the prisoners, "the commissars cannot understand your resistance. We are twenty to one and have many guns. Our Commander expected to take Bereznik in three days, but the soldiers will not attack any more over the snow against your awful machine guns."

The troops at the Vaga battles could not be compared with the unruly, Bolo rabble of the early days. They shot low and were well officered by officers, mostly Letts, who had been trained in Trotsky's military schools at Moscow.

Another explanation might have been in the story of some of the prisoners, but which was never confirmed, that the soldiers had met in a solemn, protest meeting, following the last costly, Vaga offensive, and shot their Commander for his persistence in pushing on, despite the heavy casualties. The fatal potion of Kerensky's Order still poisoned the blood of the Russian army, and although the Soviet soldiers gave exhibition of great bravery, and were well led, they were not great soldiers; they failed in the ultimate trial, and did not go through to victory when stamina and resilience for the last lap would have won.

As the Vaga men had gone furthest in fulfillment of a vain and futile mission, had parried the heart thrust, and beat back its violence, so were they the last to leave, and were still in battle at Malo Bereznik at the close of May, six months after the Armistice, that proclaimed Peace to an afflicted World, and poured cooling balm on a million wounds, so far from feverish, strife torn Russia.

Not until June did they meet their regimental comrades, coming from every compass point of the wide province, save the seabound, impassable north, to assemble at Economia for the homegoing. There the battles of Kodish and the Railway, Onega, the Vaga and Dvina and Pinega Valley were fought again, until the white, Russian snows were hued rose red with blood of recounted slain, until American soldiers sailed away, bewildered still at this gambling murder game, and sacred life—the most contemptible stake in the mad lottery.

Not the Vaga men to idly speculate on causes! They knew full well the colonel's words, and were exalted still by the fervor of their sacrificial avowal, the noblest of mankind—to lay down life for a friend.[39]

---

39   John 15:13 "Greater love hath no man than this, that a man lay down his life for his friends."

When the snow mounted high the fortifications had been made safe against any projectile
save a six-inch shell

*Trenches near an outpost at Verst 444. 339th Infantry., 85th Div.,
Vologda Railway Front*

PINEGA

19th March, 1919.

C. G. Tours.

HQ: 3407, Following telegram repeated from Archangel quote Information as to future possible relief for this expedition would materially improve the morale of troops after their long winter of Field Service, and it would also assist me in making arrangements for the future. So far I have not received any official information as to prospects.

Signed Stewart unquote.

Repeated to G. H. Q. and Agware.

Wheeler.

"It has always been a cardinal axiom of the Allied and Associated Powers to avoid interference in the internal affairs of Russia. Their original intervention was made for the sole purpose of assisting those elements in Russia which wanted to continue the struggle against German autocracy, and to free their country from German rule, and in order to rescue the Czechoslovaks from the danger of annihilation at the hand of the Bolshevik forces."

G. CLEMENCEAU.

D. LLOYD GEORGE. WOODROW WILSON.

V. E. ORLANDO.

SAIONJI.

From note, dated 26th May, 1919, Allied and Associated Powers to Admiral Kolchak.

PINEGA

THE Orthodox Church of Russia is hated by the Soviets with an intense and vehement hatred, for the institution of kings was sustained by religion even more effectively than by the Imperial Guards. Therefore, no opportunity to deride reverend personages and sacred objects is ever neglected by the Bolsheviks, or to violate with leering and uncouth pleasure, the hallowed worship places.

Under the nimbose influence of Red Moscow, the religious precepts of the people will be snatched ruthlessly from them. Harsh and unyielding though these precepts be, they are the only note of spirituality in the life of the moujik, and without them he wallows in a mire of crass animalism. There was in Holy Russia many a homily in patience and honesty and humility; but will these homely virtues endure in the arid waste and the spiritless air of agnosticism?

At Pinega, some ninety miles east of Archangel (and nearly one hundred fifty on the devious road), the cleric party was well fortified, and the outstanding civic feature of the city was the ancient monastery, standing commandingly at the edge of Lake Soyla.

The Pinega monks were quite naturally opposed to the Bolsheviks, but the mayor was a Soviet, and the city was divided in allegiance between White Archangel and Red Moscow when the detachment of Americans came in October.

The Americans' presence shepherded the wavering ones to the fold. A company of Home Guards was organized, and from outward signs the cause of the Allies had ascended to triumph. But the surrounding Bolsheviks were far from disbanded. They gathered in much strength under the leadership of Kulikoff, a competent horsethief, and commenced to plunder the slender, household larders of the peasants in the lower Pinega valley, to whose succor a police force of thirty-five Americans and two hundred White Russians were dispatched in mid-November. This police party penetrated eighty miles southeast and took Karpagora, after an engagement, but early in December was overpowered by the returning Bolsheviks. A few of the Americans were killed, more wounded, and the rest went back to Pinega, posting the White Russians in outlying villages as they retired.

So critical was the outlook that another American detachment came the one hundred and fifty miles from Archangel, ten days' journey in the darkness and the cold. But, more important to Pinega than these Christmas reinforcements, was Joel R. Moore, who came with them, wearing the shoulder straps of an infantry captain for the time in being, but whose life profession was that of college instruction, as skilled in applied humanity as the classical Humanities, and possessed of tact and understanding and sympathy, and that indefinable gift of leadership. He organised the Russians for their own defense in this bloody internecine fight, and shamed their leaders to vivid consciousness of dreadful responsibility to their pitifully dependent people.

In February, a vicious and prolonged attack in conjunction with the great Vaga offensive was made on Pinega, but the defense was well held, and when the situation looked most strained, and the fall of the city almost sure, the Bolsheviks slackened and fell back without overt cause or reason for relenting in their fierce assault, just as they did on the Vaga when the life of the Expedition was the stake.

No soldier who was in it will ever forget that mid-winter march from Archangel in gray days and cold, when the spruce trees cracked in the frost with the report of rifle shots; when the wind, a blearing blast, swept down and piled great billowy swells on the whitened trail, covered men head and foot like powdered, clownish figures, plastered their eyelids and nostrils grotesquely white with hoary frost, and flicked snow particles under headgear where they stung with the sting of pelting sand; other days when oppressive calm would stifle the air with the mystery of eternal stillness, jarringly profaned by the crunch of heavy, marching feet, the shambling of the little convoy ponies; and the tenacious trail would lower to great sheeted space, that swelled to the summit of long hills where village roofs were etched in steel on a burnished background, where the ineffectual sun strove vainly to thrust back imprisoning cloud curtains, slate hued and black.

Sometimes the way brought the soldiers through the phantom glade of a fairy forest, where delicately spun aigrettes and fragile, filmy plumes held by doubtful tenure on a limb would wave precariously in the wind and be lost in shapeless, irretrievable chaos of crumpled snow, but tens of thousands of others would fill their places, and inconceivable, bizarre festoons would spring to magic life, countless balloons and garlands and wreaths, and massive, ponderous globes, all shaped by the infinite artistry of the frost in an endless profusion of enchanting wonderment.

Sometimes their canopy would be a lilac sea, with islands of suave saffron, and slender, garish emerald reefs, which could never escape the tristful quality of the haunting Russian skies, where tragedy and melodrama ever unfolded till night clasped in blackness the brief twilight of those doleful winter days.

Under their humble roofs, the patient people revealed a hospitality that was moving in its utter absence of guile. The cherished samovar would be brought forth from a covert trove to kindle the uninvited guests with steaming tea, and in the evening all the villagers would troop to the crowded huts to doff their hats and cross themselves with pious orisons, and gaze with never wearying gaze at the strangers from the far fabled land of miracle and hope. Years from now moujik grandmothers will group rapt children around the oven stoves to tell them of the strange Americanskis who once came so many miles in the dread winter cold to help afflicted Russia.

Out in the frigid night, the aurora of the north swung swaying evanescent curtains, now fluttering with faint ethereal light, now springing to flowing, colorful life again, and one could fancy that Thalia[40] signaled from the night heavens a playful spectral heliograph, mocking these silly little men so far below, that strove to conquer the dread elements of that gaunt Northland.

But, if in the whole campaign the somber veil of tragedy was ever lifted, it was at this front where the altruistic intention of the Allies seemed to have caught the consciousness of the people (whether or not this intent was in fact altruistic), they bore not only benevolence, but even humble touching gratitude towards their deliverers, and even, undertook the burden of their own battles. Many Russians were lost in these battles for Pinega, but after the first expeditionary engagements not one American fell.

In January there was a massed assault, and when the fall of the city seemed almost sure, the Bolsheviks slackened and fell back, with their blade poised for the heart thrust.

But in March the defenses were safe in the competent hands of a regiment of White Russians, who were the de-fenders of their own towns, and the "Allied Legion" of no nation. Likewise there

---

40   Θάλεια or Thalia, "the joyous, the flourishing." The Eighth of the Nine Muses. The Muse of Comedy.

were two field guns with a Russian personnel of artillery, a unit of Russian machine gunners, carefully trained in the service of these rapid, death-dealing instruments of specialized modern war, and all these soldiers of Russia raised their heads high and proud as eagles, wearing no man's collar.

So it came that the Americans were free to take their leave for more pressing fronts and were given "Farewell and come again" from the hearts of the Pinega people, with generous, overflowing good will, abounding grateful acknowledgment of their genuine, upbuilding service. Perhaps this was more the conceived purpose of the Expedition to sustain the foundling democracy of Russia, to strengthen and instill solidarity and faith in the hearts and counsels of the Russian people, and to achieve such end by unsanguinary means. Perhaps the means might have been different and the melodrama never enacted if a college professor, with methods of applied humanity, had directed from the outset. But it is to offend the military to consider thus, and to be guilty of shameful heterodoxy.

Patrols were often clad in white smocks

RETREAT

"There is no use people raising prejudice against this expedition. Every one knows why it was sent. It was sent as part of our operations against Germany. It was vitally necessary to take every measure in regard to Russia during the war which would keep as many German troops as possible on the Russian front, and reduce that formidable movement of the German armies which carried more than a million men to the Western Front, and which culminated in that immense series of battles which began on the 21st March last year (1918)."

WINSTON CHURCHILL, Secretary of State for War, in the House of Commons, 3rd March, 1919.

159489   Brig. Gen. W. P. Richardson about to cross the Dwina by reindeer team. Gen. Richardson arrived in Archangel shortly before the last of the few remaining Laplanders drove their reindeer teams northward. Archangel, Russia, Apr. 17, 1919.

RETREAT

WHEN the appeal to patriotism failed, Archangel Province, under British direction, invoked conscription, and by the middle of June, twenty-two thousand Russian soldiers had been assembled by coercive means.

They thronged the backward villages through which the Americans passed on their way to disembarkation, and looked very fresh, like college youths, as they sauntered up and down to an eternal serenade of wheezing accordions, or with sacerdotal, marching chants, went swinging by in platoons and companies, these young conscripts, who knew so little of war and its harrowing disillusionment.

For the moment all breasts were filled with that contagious ardor that springs from every massed effort, no matter its end, but not one in a hundred knew or felt the call of patriotism for the coming conflict of Russian against Russian.

There was cause enough for the fight had it only been revealed to these pliant, guileless, peasant folk. For their country, weakened, helpless and faint from many war wounds, was being debased by vile and vicious poltroons who had stamped out the holy fires of the Revolution, nullified the Constituent Assembly, and stifled every voice of liberty with hands more remorseless than the cruel manacles of the Tsars.

The cause was there, but if their mentors sensed it, they manifested almost incredible obtuseness in failure to impart these moving eloquent reasons for the fight. They were silent about the odious exploitation of the masses un-der the crafty, artful guise of proletarianism; they said nothing of the wicked violation of sacred property rights, the unprincipled plundering, the trampling down by power maddened feet on all revolutionary enlightenment, the desecration of all things spiritual, the wanton derision of the church which had been the faith of the people and of their venerated, sainted fathers.

Here was reason enough for any Russian with exalted, holy devotion to lay down his life for his stricken country. But instead of such scathing and unequivocal indictment, the British dwelt upon the conduct of the Bolsheviks, shameful and faithless towards the Czecho-Slovaks, and gave out, with venomous vituperation, highly colored stories of enemy atrocities and cruel treatment of prisoners so patently over-extended that they failed to make a convincing impression even on the moujik mind.

So soon as navigation opened, there commenced an exodus of Russian officers to Archangel, sent by the British Command to lead the newly formed native legions. These officers came from the old Imperial Army, many were titled, proud of their high birth, and by every thought and training, and by every instinct, irreconcilably opposed to every notion of social equality; in short, irredentists of that heartless, arrogant, military class which a worn afflicted world had cast off in a travail of four years' agony and afflicting grief, and long suffering Russia had driven forever from her temples.

So the fresh formed conscript ranks were made conveniently vulnerable for Bolshevik propaganda, this new weapon of warfare, invisible and treacherous, that on the Eastern Front had scored such havoc with the boasted discipline of the Germans. Soviet agents were everywhere, mingling with the people on the streets of Archangel, wearing the khaki of the newly organized soldiers, living with them, going through their drills, and fatigue and exercises, and ever with the passionate zeal of fanatics, feeding them the poisonous doctrines of Reddest Moscow, ceaselessly, night and day.

Now the innuendo was very plausible that these aristocrats of the Old School had returned to restore the Romanoffs, and that the British capitalists were leagued with them for the conquest of Russia and the enslavement of the common people. It was easy to argue that the British, always interested in the trading possibilities of Archangel, had come to exploit its resources. Otherwise why should they be so vitally concerned in this civil war of Russians? British officers were freely mingled with these Imperial officers, British Intelligence supervising the staff work and dispositions, and a liberal spreading of reliable British N.C.O.'s among the ranks, to keep a watchful eye on things and bolster the recruits in the stern trial of first battle.

The great majority of the British officers had no appetite for the business ahead. They were tired and homesick, weary and fed up with war for all time after four racking years of it. Moreover, they disliked everything Russian with a withering aversion, and in their forced association with the Russians, treated them with a disdainful condescension and that impersonal, inhuman lack of tolerance which is British beyond all imitation. Openly they distrusted their allied comrades, and sometimes when tired and irritable and nerve frayed, they said so, which did not make towards the establishment of an enthusiastic and permanent entente, for the educated Slav is an accomplished linguist, and sometimes he understood and did not easily forget when he was abused in English, and vehemently cursed as a "bloody Bolo."

It had been determined before the opening of navigation that all American forces should be withdrawn and the campaign abandoned. The reason for this was not revealed to the troops just as the cause of the Expedition had never been mentioned, and every man in American uniform sensed a gaping moral void on the part of his Country. Certain death from the Bolsheviks awaited those loyal Russians who had placed their trust in the promised salvation of the Allied leaders and the American authorities at least seemed blind to their manifest duty to the Archangel government. It was an awkward situation for the statesmen, but unavoidable under the circumstances—and Archangel was a long distance removed from Washington. Anyway, the British held on—they would have to attend to uncomfortable details. We were going to clear out, and clear out we did.

The problem of evacuation was a disturbing one. There was a clamor in England as insistent as that which echoed from America to get out of Russia and get out without delay. This might have been done, and the British might have abandoned these thousands of Russian people who, trusting in the courage, the steadfastness, and the honor of the Allies, had cast their lot with them for better or for worse. But, instead of deserting the country without ceremony as we did, a frank disclosure of the situation was made to the press in England, and a call was issued for volunteers to rescue British soldiers at Archangel. A mixed brigade of venturesome men who were wearied by peace time tedium and longed again for the thrill of war, and others who were out of work and could get no other employment, was raised by this method, but to muster the full quota for relief it was necessary to add a like number of Regulars, in all approximately eight thousand men. Each brigade had two infantry battalions, units of artillery, airplanes, machine gun corps and engineers, and the first echelon, commanded by Brigadier General G. W. Grogan,[41] Victoria Cross, reached Archangel at

---

41   George Grogan (1875-1962) attended Sandhurst. He fought in Sierra Leone in the Hut Tax War of 1898. He served four years on the western front, earning several decorations. Postwar, he served as aide-de-camp to King George V.

the end of May. The rest, under Brigadier L. W. Sadleir Jackson,[42] came on the 10th June, and the ships that brought them carried away the Americans.

To the civil mind an evacuation, especially by sea, seems a simple matter. The civilian thinks of it merely as a packing off to the ships, disregarding the losses involved to make short shift and get away. But in complicated, modern war, there are countless perplexing details in the final movement of an army. Massive, ponderous ordnance and munitions and supplies must be assembled with prodigious labor, transported or destroyed. And it is necessary to hold the enemy off till the last retreating file has mounted the gang plank and put off far to sea. Also, in the case of Archangel, it was an involved problem to attend to the civilian population.

The British government laid open the offer to transport every Archangel resident apprehensive of the Bolsheviks, and to provide employment for them in other lands. It was expected that vast numbers would avail themselves of this opportunity and would flee from the approaching reign of horrors, but when the time came only sixty-five hundred and thirty-five came forward for expatriation, and these were all sent to South Russia and the Baltic States.

When all was in readiness, General Ironside planned to safeguard the retreat by administering a sharp "disengaging blow," like Sir John Moore dealt the French at Corunna one hundred years before, which would shake the enemy's morale and disabuse him of any notion of following the retreating troops to the waterside.

The Czechs had fused with Admiral Kolchak's armies. Under the leadership of General Gaida, they formed his right wing and were beyond Perm, some three hundred miles east of Viatka. It was thought that these friendly Siberian forces could take Viatka, advance up the railway to Kotlas, and join there with the Archangel Russians. Thereupon the British, leisurely and in security, could return down the river to the waiting transports and sail homeward.

So Kotlas, which had been the original objective of the River Column, became the objective once more. The Admiralty dispatched to Archangel a flotilla of gunboats, monitors, mine sweepers and many other craft for the transportation of troops and supplies to act as auxiliaries for the infantry, and again the Dvina

---

42   Lionel Sadleir-Jackson (1876-1932) served in the Second Boer War. He spent four years on the Western Front. He earned several decorations and awards. After the war he oversaw troop training in the British Mandate.

became a scene of skeltering preparations for war.

On the 2oth June, the disengaging offensive began; the British and Archangel troops attacked across the river from the Allied position at Toulgas, and gained complete victory, capturing two hundred prisoners, many machine guns and three field guns. But now word came from the south that the Bolsheviks there had concentrated in great forces against Kolchak and had utterly routed him, that he was fleeing east, had already retired as far as Yetakerinburg, and all hope would have to be given up of effecting a junction with the Siberian army.

So the importance of taking Kotlas waned, but even if Kolchak had not failed the advance could have gone little further, for it was found that due to the light snowfall of the previous winter, the waters of Dvina were low, beyond all precedent, and the British flotilla could follow no farther upstream.

Most discouraging of all, treachery broke out in all quarters from the allied Russian troops. On the 7th July a battalion held in reserve on the river mutinied in the night and murdered three British and four Russian officers as they slept; four other officers were seriously wounded. On the 22nd July the whole Onega detachment went over to the Bolsheviks, and the safety of Archangel became seriously jeopardized from this west port. Nearly at the same time British firing squads suppressed a revolt on the Railway front before the Russian mutineers gained the upper hand.

Many of the British officers had passed through all the harrowing fires of France, but here was a form of peril new in the experience of the most hardened ones—base betrayal by the sentinel who kept the black watches of the night, and treachery in the heart of the citadel from hands stretched forth in friendship. The brave man, standing on his feet and facing the end, does not fear advancing death; but now it lurked in hiding, it descended in the night and struck from the dark upon unconscious sleep, so that tired soldiers dared not rest, and the strain snapped nerves of steel.

A few weeks before these outrages, Toulgas was given over to a defense that was entirely Russian. Shortly afterwards, in the uncertain light of early morning, on the 25th April, there was a wild commotion, and, following interminable confused firing that sounded from all quarters of the village streets, a lamp message flashed across the Dvina to the Allied position at Kurgoman: "We are completely surrounded; the Bolos are attacking in five places."

Shortly thereafter, through a fusillade of bullets, a Russian officer, with two men, effected a passage of the river in a small boat, and told the shameful story of how nine officers had been murdered as they slept and bloody Toulgas delivered by faithless Russian soldiers to the waiting Bolsheviks in the woods. Through a prodigy of bravery by a handful of loyal artillery men, the guns were pulled back to Shusiga, ten miles downstream, but it was not until the middle of May that Toulgas was retaken, and while it stayed in enemy hands, the Allied position was alarmingly critical with the right flank over the Dvina completely turned.

Thus, with mutiny breaking out in all quarters, the virulent propaganda of the Bolsheviks bore malignant fruit beyond their most sanguine hopes, and the situation was menacing enough to alarm the most conservative in Allied Councils. Had it not been for the two splendid reinforcing brigades, the often imperiled life of the Expedition would have been destroyed at last. The British War Office for once became thoroughly apprehensive. General Lord Rawlinson was sent to preside over the leavetaking, and fresh reinforcements, two battalions of infantry, two machine gun companies, two batteries of Royal Field Artillery, one engineer company, and five tanks were rushed to Archangel from England.

The intention had been to complete the evacuation just before the closing of navigation in late October, but now it was seen that this might be too late, and in the present urgency no time could be lost. "The disengaging blow" was delivered on the 10th August by Jackson's sterling brigade, a little beyond Seltzo, the furthermost south achieved on the Dvina by the little River Column almost a year before. Two thousand prisoners were captured, eighteen guns and many machine guns, and the rout was complete. With the enemy now safely at bay, the British turned the defenses over to the Archangel authorities, who persisted in staying, although they were advised that it was suicidal to do so, and "friendly intervention" was brought to an inglorious, albeit an unbloody, close on the 27th September, eleven months after the Armistice that had outlawed the rule of warring strife as the arbitrament of discordant nations.

When the last troop ships trailed off to drooping skies, a bearded moujik sat in the stern of a flat boat directing four broad backed women at the oars. The recumbent coxswain waved a languid gesture across Archangel Bay where tiny ships were bearing off to the north; and four oars poised in mid-air as the laboring crew turned with dull Slavic contemplation to regard the parting foreigners, and the end of their peculiar expedition. But only for a moment, there was more important business in hand than idle gazing at Englishskis, however queer they might be. A gruff command, and the freighted craft continued its slow toiling course to the market place, the overlord resumed his interrupted smoke of good Allied cigarettes and the Englishskis were dismissed from memory. This was the leave-taking.

On the evening of 12th October, 1919, the last of the Allied forces set sail from Murmansk for England; four months afterward, on the 20th February, the Bolsheviks recaptured Archangel.

Nearly four months earlier the last of the Americans set sail on the 26th day of June, 1919, and as the paling shores mingled with the distant sky line and faded from sight, so too the fever of this troublous, little war with Russia abated, yielding to the gentle ministrations of memory's cooling twilight.

With the Americans, at least, there remained no shred of illusion. When Winston Churchill told the Commons that Archangel, with one lone American regiment, the few battle retrieved soldiers of England, and a single battalion of disaffected Frenchmen, had kept many German divisions in the East, and played an important part in the last battles, he laid a flattering unction to the soul of British statescraft; but his insincere words did not deceive the American soldier, for the American soldier was mentally and emotionally paralyzed beyond deception, and a conviction of blunder was only strengthened by this and other clumsy explanations vouchsafed by Allied statesmen; by the guilt laden silence of America.

Germany was never concerned with Archangel. There was no evidence of German participation in the campaign; no evidence that our petty hostilities with the Bolsheviks had ever benefited Foch on the Western theater.

We had waged war upon Russia. Whether willfully or unwillingly, our country had engaged in an unprovoked intensive, inglorious, little armed conflict which had ended in disaster and disgrace. Perhaps this was a laudable thing to do. Perhaps it is always idealistic and praiseworthy to intervene for self-conceived righteousness in the internal affairs of another nation, as England might have done in the case of the American Confederacy, and as we did in the case of this civil war among the Russians. It is easy enough to enter the battle lists, but, once in, it is not so easy to withdraw from the fight with self-respect unsullied and honor undefiled.

So Archangel proved, with its sullied record to blight forever the good name of America when soldiers gather to tell of the Great War, and, great as the cost of the campaign had been with 2,485 casualties[43] of killed and wounded and sickened men, its financial loss, over ten times the price paid Russia for the vast dominions of Alaska, there was not a man in the ranks who did not sense the disgrace in our ignoble desertion, there was not an American officer who would not have chosen to have left his bones bleaching white beneath Archangel snows, than been a living witness to the ignominious way in which his country quit and slunk away.

All felt a personal sense of poignant shame for the failure to see the game through to its uttermost bitter end, or else seek expiation by honest avowal of wrong and humble contrition. It was an inexorable dilemma, one that took the staunchest courage, no matter which course was followed. Perhaps the higher courage would have been the admission of culpable fault. But we took neither course. We merely wilted from Archangel and came away.

On the homeward troopships, among the ice floes of the White Sea, the taunting unspoken reproach galled most bitterly of all, for we left our British allies to extricate themselves from the miserable mess as best they could, and with no explanation and never a sustaining word we left them.

Many trying things in the campaign had aroused the Americans to intemperate speech, which now to recall they would have surrendered all they possessed. Incompetence and tactlessness, and seeming lack of understanding and sympathy by those in power, to which the soldiers of England appeared indifferent, never failed to draw the intense, iconoclastic fire of the Americans. The difference lay in the national atmosphere of the two countries, the divergence in character and traditions, born and nurtured under the republican and the older order. They are a different people from us, the British, though the blood strain be the same. The glory of baseball is lost on them; they play the tedious cricket; but, when the fight is on, the quality of the bulldog, once at grips to hang on with set teeth till death, is British; blinded to all save the solid grimness of the task in hand, their brains seem dull to those imaginative flights which are the curse of the Western soldier.

---

43   Author's footnote: from the Chief Surgeon's Report.

Thus ended America's share of the war with Russia. At Brest the "mutinous" regiment was shunted in fragments over the seas to America, and in the homeland, these soldiers who had borne arms in conflict six months after the Armistice, were shooed off to civilian life, and the whole embarrassing matter was expunged from the war record.

All inquiry concerning the Expedition has been met by specious pleas in evasive avoidance. No peace was ever made with Russia, as no state of war had ever been recognised, and the legalists might well contend that all who engaged in it are open to indictment for manslaughter, for the enterprise will always remain a depraved one with status of a freebooters' excursion.

At Corbela sat an aged woman with ghastly face, gray as the dirty platok[44] that framed it, her gaunt chin resting on a hand, bony and hideous from relentless toil. With failing despairing eyes, she saw in the dwindling snows only the dissolution of winter, quite blinded to buoyant spring that with tufts of brown turf bursts boisterously through the southern hill slopes, like heedless youth that with surging, eager, passionate desire presses on the reluctant heels of death to life's fulfillment.

Outside the hut a young moujik, with the handsome physique of first unsullied manhood, and the credulous eyes of a child, curiously watching the north marching Americans; a giant of masked strength, needing only the key of trained intelligence to unloose immeasurable dynamic force that might some day rule the world.

Kindle the liberating torch of enlightenment in the nether regions of the Slavs, strike from the millions the shackles of serfdom ignorance, and from the pestilential ashes of present degrading Bolshevism, Russia, the giant, in stupendous power, rises phoenix-like to Jupiter.

To the Russian people we owe a debt that can never be paid except in deepest and very humble gratitude; for, when those gray hosts swept over Belgium and Northern France, Russia invaded Prussia, threatened the gates of Koenigsberg, routed the Austrians in a smashing blow at Lemberg, and, when the German aggressive movement was at its culminating height, drew off to the east two Army Corps and a Cavalry Division from von Kluck's right wing, a fatal diversion of the German forces which enabled Joffre, closing in the breach at the Marne, to save Paris and turn the advance into

---

44   Traditional Russia shawl or headscarf.

a complete retirement.

This great battle of the Marne marked the initial phase of the war, and completely frustrated the cherished Berlin plan of gaining quick victory by tactics of overwhelming surprise.

Many anxious months followed as England slowly transformed her energies from peaceful pursuits to those of war, and during this prolonged, crucial time the Russians never wavered from the attack. They massed for repeated hammering offensives in Poland, in Masturia and east of the Vistula in Galicia, so that the German Imperial Staff could never develop full strength, but had to be content with a holding campaign in the West while marshalling most forces to oppose the menacing East.

Not until the beginning of 1916, because of the Russians, could another effort of masses be made. Then every available man was concentrated with the Crown Prince's army as he smashed at Verdun to bring France to her knees, but when the assault was at its height, again obedient to her trust, and faithful, Russia sprang to the attack with such heroism and such devoted and reckless courage, that the controlling German combat divisions which might have gained the fortress had to be diverted from Verdun to Galicia.

Yet again at the commencement of 1917, at Mitau, and, in the summer of that year, when the British Empire assembled its legions at the Somme, Brussiloff struck south to the Carpathian passes, and it was only when Russia collapsed exhausted, and ghoulish Bolshevism looted the prostrate stricken gladiator, that the united German armies marshalled in full strength for a crushing blow. *Only then did Germany have numerical superiority in the West.*

We can gain an impression of what might have happened from the fury of that La Fère-Arras offensive, which shocked the world by its blighting trail of spectral horrors; hardly a British Division was left intact, and France reeled and staggered in a nausea of mortal weakness until Clemenceau in agony cried out to the Allies for sustaining support.

All might have ended then, had it not been for America, but America could never have come, had it not been for the Russian sacrifice in the early days, when the German Divisions, fresh and recklessly rash, were filled with the lust of battle conquest, and the German leaders, careless of casualties, flung their men to death with a high and free hand.

It is well to remember these things when we boast (a little noisily) that American arms won the great war. No one nation won this appalling contest of the nations embattled at Esdraelon,[45] and, great as our offering was, how small it was and how feebly comparable to that of Russia who laid down the lives of more men than all we sent to France, and paid a ghastly toll in crippled, maimed and battle losses, a million souls beyond the sum of our whole military effort!

A joint Resolution, providing for any needed explanations and reparations which may be due from this country for our invasion of Russian territory was introduced in the United States Senate at the second session Sixty-sixth Congress by Senator France, 27th February, 1920.

February 27, 1919. Funeral of Wagoner James E. Byles, Supply Co. 339th Inf. 85th Div. Casket covered by American flag being lowered into the ground. Chaplain John S. Landowski presides over the service.

161680

45　　　　Esdraelon refers to the Jezreel valley, or the valley of Megiddo, scene of several battles from the Old Testament. The word Armageddon is derived from this. The Battle of Megiddo described in the Book of Revelation is the penultimate battle between good and evil. Revelation 6:16 (BSB) "And they assembled the kings in the place that in Hebrew is called Armageddon."

APPENDIX A

Trotsky's Note on the Allied Troops in Murmansk

The measures taken by the People's Commissariat for Military Affairs to deal with the landing by our former allies at Murmansk are completely in accordance with the instructions I received from the Council of People's Commissars and, in particular, from the Commissariat for Foreign Affairs.

Any attempt made by our former allies to transform the White Sea coast into a base for their operations will meet with an uncompromising rebuff from us.

As is known, I have dispatched the armed forces needed to safeguard the Northern coast against any encroachments whatsoever.

The force landed by our ex-allies is numerically insignificant, and more symbolic than effective. The Anglo-French imperialists apparently count on establishing in the North a pole of attraction for all sorts of adventurers, mercenaries, counter-revolutionaries and traitors. To this end, our ex-allies have long since been bribing certain groups of the White-Sea coast inhabitants, and, in particular, the Murmansk Soviet and some of the military and naval representatives in the area.

At the same time, an attempt was made by French and other officers to move to the North substantial units of Czechoslovak, Serbian, French and Russian White Guards, especially airmen, so as to form a powerful occupying force at Murmansk, and later at Archangel.

Two groups of prisoners-of-war, consisting of 100 Serbs and

200 Italians, did actually succeed in getting through to Archangel, with a certain quantity of weapons. A most searching inquiry is now under way to establish the routes by which these groups traveled and who it was that helped them.

In accordance with my orders, these two groups have, of course, already been disarmed and placed under arrest.

The central food-supply administration received an application from the French military mission for an issue of foodstuffs for a thousand men who were allegedly being sent through Murmansk to France. This is, as we know, the formula by means of which adventurers, mercenaries and crooks are being mobilized for the occupying forces. Officially, they are being sent 'to France', but in reality they are destined to raise a revolt on Russia soil and to seize our Northern coast.

A few days ago a group such as this, consisting of a few dozen Czechoslovak and Polish White Guards and French officers, was detained in Moscow and put in prison. The measures taken provide some guarantee that no further sudden movement and concentration towards the North by similar groups can occur. Those Russian traitors who treat as normal the barefaced arbitrariness committed by foreigners in our North, and provide help to it, will be dealt with in short order.

The picture before us is now extremely instructive for any honest observer. Exactly the same groups and classes of the population show themselves Anglophil or Germanophil in orientation, depending on whose help is nearest to hand. The Cadets and Right SRs go along with the Japanese in the Far East, in the North with the British and French, in the Ukraine and on the Don, and at Pskov and Dvinsk, with the Germans, and the Cadet who makes an agreement with Skoropadsky in no way blames as unpatriotic the Cadet who is ready to sell Russia to the Anglo-French stock-exchange speculators, while the latter fully 'understands' his colleague in the Ukraine.

Krasnov operates according to a German orientation. His brother Dutov leans towards the Czechoslovaks and the British. The third man, Semyonov, has hired himself out to Japan. All three of them are fulfilling the instructions of the Russian bourgeoisie. This is their patriotism, their national dignity, their national honor.

In conclusion, I should just like to draw attention to the specific activity of the French military mission in Russia during the revolution. It is hard to conceive anything more limited, short-sighted and helpless than a French petty-bourgeois clad in a General's uniform or a diplomat's frock-coat. Above all, this petty-bourgeois is ignorant of geography and incapable of finding his feet in an unfamiliar setting. As a result, the activity of France's agents in Russia was entirely directed against the elementary interests of France. I shall not deal in detail with the actions of the French diplomatic and military representatives, but will mention only the most important of these.

France raised up the Romanians against us—and the Romanians ended by transporting the German troops into New Russia.

The French raised up the Rada against us, helping it with money and military leadership – and the Rada ended by allying itself with Germany and Austria-Hungary.

The French supported Kornilov, Kaledin and Krasnov—and Krasnov is working with Skoropadsky.

It was the French who pressed hardest for Japanese intervention. But one would need to be really as innocent as Tartarin to suppose that Japan wants to get involved in an armed conflict with Germany, and not merely to grab the Russian provinces of the Far East.

This was, and still is, the policy of all the agents of France on Russian territory. Mr. Clemenceau is nothing but an hysterical petty-bourgeois, a journalist who has not recovered from a state of chauvinist intoxication. He is in charge of the policy of unfortunate France, which has been drained of blood. Through his agents he is everywhere creating enemies for himself.

Let us actually try, in a calm way, to answer the question: what is it that the British and the French want? They want to involve Russia in the war, to create a new Eastern Front. The Soviet power does not want this. Hence the idea of overthrowing the Soviet power.

Let us assume for a moment that they succeed in their aim. Does any sensible person imagine that the working class and the revolutionary poor peasants, who undividedly follow us, would quietly and for a long time put up with the establishment of bourgeois government that made an alliance with Anglo French imperialism?

The moment that the Soviet power was overthrown would

see the beginning of a civil war throughout the country on a scale twice and three times as great as before. There could be no question of Russia making any contribution to the war under these conditions.

A Russian bourgeois government would find itself under such pressure from the working people that any independent policy would be quite beyond its capacity. A government headed by Milyukov and Kerensky in Russia would be incomparable weaker even than Skoropadsky's government in the Ukraine. And Skoropadsky's government depends entirely on the support from foreign bayonets.

In the immediate future we shall extend this mobilization of certain age-groups to all parts of Russia.

I do not doubt that the All-Russia Congress of Soviets will sanction the transition to compulsory military service for the sake of protecting the security of the Soviet Republic from imperialist onslaughts. And then the last word on all this will be spoken by the working class of Europe and of the world.

APPENDIX B

Petition to Bring Troops Home

PETITION TO
CONGRESS OF THE UNITED STATES OF AMERICA

FIRST, in making this our petition, we do hereby affirm our unswerving loyalty to the Country and Government of the United States of America, and do hereby express our willingness to abide by the acts of the constituted authorities should this petition be either tabled or denied. This petition we make in manner following:

BELIEVING from the sifting of evidence that the American Units in North Russia not only are suffering incredible hardships, but are in grave danger at the hands of an overwhelming and conscienceless enemy, and

ALSO BELIEVING that the Archangel expedition, if it ever had a valid excuse, cannot now be justified, neither on the grounds of humanity nor of military expedience, and

ALSO BELIEVING that now the war is practically if not technically over there exists no patriotic reason why our American soldiers in North Russia should not have at least an equal chance for their lives with other American soldiers,

WE, THE UNDERSIGNED, THEREFORE, RESPECTFULLY PETITION for the withdrawl of the American Soldiers from the entire country of North Russia and their return to the authority of their own officers and the War Department of the United States of America; or failing this, we urge that they be properly housed, fed, clothed, munitioned, and given proper hospital facilities and reinforcements without delay.

FURTHERMORE, in appending our signatures to this petition we do jointly and severally declare in the most solemn manner that the we have no political party to serve in the premises, but that we do make and constitute this our petition with the sole desire of releasing our American soldiers from an ambiguous, intolerable, and entirely un-American situation in which death is the least among many evils.

Dated at Detroit, Michigan, this _____ day of _____ in the year of our Lord 1919.

NAME	ADDRESS	DATE
Mrs Frank Peters	Elizabeth New Jersey	2/4/19
Mrs Paul Lanckriet	108 Louis ave	2/4/19
Mrs Rose Dequith	98 Louis ae	2/4/19
Mrs Desire Lanckriet	98 Louis	2/4/19
Mrs Victor Deneweth	660 Belinede ave	2/4/19
Mrs Alvin Vanden Houge	2098 Harper ave	2/4/19
Mrs Fred Lanckriet	98 Louis ave	2/4/19
Mrs Irene Hallaert	98 Louise ae	2/4/19
Mrs Kenda Kemper	98 Louis ave	2/4/19
Mrs Van Gale	697 Belived	2/4/19
Mrs August Lanckriet	697 Belived ae	2/4/19
Mrs Charles Lanckriet	2098 Harper au	
Mrs Joe Vantooks	950 Belinide	2/4/19
Mrs May Vanoyhey	2190 Harper ave	2/4/19
Mrs Emile Vanden Hagen	French Road	2/4/19
Mrs Peter Demain	St Jean and Gabriel	2/4/19
Mrs Henry Carveny	2108 Harper ave	2/4/19
Mrs Joe Kestlas	98 Louis	2/4/19
Mrs Edmond Massial	109 Louis ave	2/4/19
Mrs Odil VanderBan	109 Louis ave	2/4/19

FIRST, in making this our petition, we do hereby affirm our unswerving loyalty to the Country and Government of the United States of America, and do hereby express our willingness to abide by the acts of the constituted authorities should this petition be either tabled or denied. This petition we make in manner following:

BELIEVING from the sifting of evidence that the American Units in North Russia not only are suffering incredible hardships, but are in grave danger at the hands of an overwhelming and

ALSO BELIEVING that the Archangel expedition, if it ever had a valid excuse, cannot now be justified, neither on the grounds of humanity nor of military experience, and

ALSO BELIEVING, that now the war is practically if not technically over there exists no patriotic reason why our American soldiers in North Russia should not have at least an equal chance for their lives with other American soldiers.

WE THE UNDERSIGNED, THEREFORE, RESPECTFULLY PETITION for the withdrawal of the American Soldiers from the entire country of North Russia and their return to the authority of their own officers and the War Department of the United States of Russia; or failing this, we urge that they be properly housed, fed, clothed, munitioned, and given proper hospital facilities and reinforcements without delay.

FURTHERMORE, in appending our signatures to this petition do we jointly and severally declare in the most solemn manner that we have no political party to serve in the premises, but that we do make and constitute this petition with sole and desire of releasing our American soldiers from an ambiguous, intolerable, and entirely un-American situation in which death is the least among many evils.

Dated at Detroit, Michigan, this ___ day of ___ in the year of our Lord 1919.

APPENDIX C

SPEECH of HON. HIRAM W. JOHNSON, OF CALIFORNIA.
BRING AMERICAN BOYS HOME FROM RUSSIA.
Wednesday, January 29, 1919

Mr. President, recently I gave a wholly unnecessary notice, but the usual one, that to-day I would address myself to the resolution which I presented some weeks ago, and which it was my expectation on this occasion to call from the table and to have considered by the Senate. The pendency of the measure, however, that is before us now precludes me from taking up the particular resolution, but, nevertheless, the time being propitious, I wish to address myself to it and to the present situation in Russia.

On the 12th day of December last I introduced a resolution asking the State Department to define to us the policy of the United States toward Russia.

I was mighty lonely then, Mr. President, in the presentation of that subject matter. To-day, sir, I have my warrant here in the letters of mothers and fathers and wives of men who are suffering and who are dying in Russia. It became obvious to me that this resolution would not be acted upon by the Foreign Relations Committee, and subsequently I introduced the resolution, of which I speak to-day, expressing the opinion of the Senate that our troops should be withdrawn from Russia. Until the introduction of my first resolution, our relations with Russia, the activities of our soldiers there, and our policy in fighting, after the conclusion of the war with Germany, with another people against whom we never had declared war, had scarcely been mentioned publicly, and the subject, apparently, was one from which the supposedly popular branch of our Government timidly shrunk. In the last month and a half, however, a very healthy discussion has been stimulated within and without this body,

and the American people have learned for the first time during this period a little, though not much, of American troops still battling on foreign soil. My original resolution was received, as I presumed it would be, in hot indignation and resentment by the part of the press which has rejoiced in its attitude of merely echoing approval of whatever the administration might do, but the subject matter has been taken up by that portion of the press of America which yet is free. Legitimate discussion has been stimulated, and our people, at last, by a modicum of free expression, have been faintly heard. What I sought to accomplish originally has in part been realized, and I think it no exaggeration to say that while the Senate, yet under the spell of the strange psychology of war, may yield its assent to a policy concerning which it was not consulted and about which it has not been informed, and which is at variance with our professions and violates our Constitution, the people of the country, emerging from this psychosis, and now far in advance of their representatives, with practical unanimity favor not only the original resolution but also the one now under discussion.

The peace conference recently has issued a formal statement concerning Russia, which, apart from its expression of esteem and love for the Russian people, asks all of the factions to meet in the Sea of Marmora on the 15th day of February. We are wholly in the dark as to what is contemplated when all parties meet upon an almost uncharted island in a distant sea; but we are becoming accustomed to the open diplomacy which in daily communiques with few words telling nothing soothe the perturbed spirits of the democratic peoples of the world. Because I assume that an address couched in such friendly language, specifically recognizing the revolution and disclaiming any desire to interfere in Russia's internal affairs, is but the initial step in a carefully prepared and thoroughly accepted plan of all the powers for the settlement of every question respecting Russia, I will not speculate upon the possibilities nor discuss the pronunciamento which has added to the case only the knowledge of the extraordinary affection in which the Russian people are held by the allies.

My interest has been in America's policy and in the soldier boys of the Republic, who, seemingly, for months were forgotten by their Government. My prime interest and concern yet are these American boys. The action of the peace conference thus far leaves these American boys where we discovered them a month and a half ago, and, so far as they are concerned, unless we take it for granted that they are to be withdrawn immediately after the 15th day of February, their situation is no different than when this subject first was broached. I ask for them nothing more than I ask for all American troops on foreign soil. I ask for us an American policy.

I am quite aware that there exist two distinct lines of policy for our country in the near future to pursue. I know that in this body the line of demarcation between these two policies is becoming increasingly plain. Recently the Republican floor leader, in an epoch-making speech, stated with clarity and precision his view of our future policy. With logical force and plainly he stated that he deemed it essential by the peace terms immediately to create the following:

1. A Jugo-Slav State.
2. A Czecho-Slav State.
3. The security of Greece.
4. Albania.
5. Montenegro.
6. Roumania.
7. Armenia.
8. Syria.
9. Palestine.
10. Poland.
11. The independence of Russia's Baltic Provinces, thus: 1, Finland; 2, Esthonia; 3, Lavonia ; 4, Courland; 5, Lithuania; 6, Ukrainia.

And, in addition, that we must take and hold ample security from Germany and Austria.

I will not quarrel with his creation of the various new nations. But the distinguished Senator prescribes our future duty in this language:

We must do our share in the occupation of German territory which will be held as security for the indemnities to be paid by Germany. We can not escape doing our part in aiding the peoples to whom we have helped to give freedom and independence in establishing themselves with ordered governments, for in no other

*way can we erect the barriers which are essential to prevent another
outbreak by Germany upon the world. We can not leave the Jugo-
Slavs, the Czecho-Slavs, and the Poles, Lithuanians, and the other
States, which we hope to see formed and inarching upon the path of
progress and development, unaided and alone.*

There is but one conclusion from the Language used, and that
is that after the creation of the 16 nations the duty devolves upon
the United States, in part at least, to maintain these nations, and
aid them in establishing themselves with ordered governments by
America's soldiers. I will not subscribe to this doctrine. I will not
for one instant concede that it is the duty of the republic to maintain order in a Jugo-Slav or a Czecho-Slav State, in Albania or
Montenegro, in Roumania or Armenia, in Syria, Palestine, Poland,
Finland, Esthonia, Lavonia, Courland, Lithuania, or Ukrainia. I
am opposed to American boys policing Europe and quelling riots
in every new nation's back yard. The perils of such a course to
the Republic seem to me obvious. Aside from the material aspect—and you may ever keep in mind that the establishment and
maintenance, in part by us, of these distinct peoples must be paid
for by our already overburdened taxpayers—I would not shed the
blood of American boys in the internal disturbances of an Esthonia, Lavonia, Courland, a Lithuania, or Ukrainia. I decline to be
frightened now by the "German menace." Germany is whipped
and thoroughly whipped. The central empires, in the language of
the President, are in process of liquidation. The German Army has
been practically demobilized; her ordnance has been relinquished;
her means of transportation delivered; her fleets surrendered; her
towns occupied. She is involved in such civil strife that it is doubtful if there will be a stable government with which a peace may be
made. She has been so thoroughly disabled that she could not, a
few days ago, resist a small advancing Polish army. She is suffering
just retribution for her offenses. The central empires are incapable
of aggressive action now. Our help, therefore, to the 16 nations
which would be created "in establishing themselves with ordered
governments" must be the aid which our arms would give to one
faction or another in these new States, and I am not ready to shed
American blood for any faction of any foreign State or in maintaining the government in any Baltic province. It is staggering to
speculate upon the number of armies and the hundreds and thousands of American boys that would be required in these 16 nations.
It is time for an American policy. Bring home American soldiers.

Rescue our own democracy. Restore its free expression. Get American business into its normal channels. Let American life, social and economic, be American again.

I heard it said recently upon the floor of the Senate that the President had a most difficult task abroad. I recognize this, and wish that some real aid could be extended to him in the one possible accomplishment which may be his. I listened with rapt attention to the scholarly and statesmanlike address of the junior Senator from Pennsylvania recently, the point of which, as I understood it, was that we should at once conclude our peace with our foes and bring our boys home, leaving to subsequent determination the formation of a league of nations. I take it that he is no more opposed to a league of nations than I am; and I am very frank to say that if the President can return to the United States with a league of nations which does not relinquish our sovereignty, and which in reality will be a preventive of future wars I shall welcome most gladly and enthusiastically, this great accomplishment.

But America wants peace. It has achieved its great primal idealistic purpose and crushed a ruthless militarism. I reecho the words of the great commander of the Union forces in the Civil War immediately upon surrender of his adversary "Let us have peace." Let us have peace and get out of Europe. On the 11th day of November the armistice was signed. Two months and a half have passed. The censored dispatches from Paris are conveying to us now the news that it is hoped the conference may conclude its labors in June and we arc being prepared, apparently, for a second trip some time in the summer by our Chief Executive to Europe. I would not minimize, of course, the magnitude of his task, but I have a very distinct recollection that much of it has already been agreed upon. Because of prior agreement among our co-belligerents the idealism of the President is likely to be severely taxed, and his difficulties will be manifold. But the pitiless logic of events, the ironclad prior understandings of our allies, have lessened his labors and reduced to a minimum the subjects with which he must deal. He announced in January, 1918, and subsequently, fourteen peace points. The first of these—

> *Open covenants of peace, openly arrived at, after which there shall be no private international understandings of any kind but diplomacy shall proceed always frankly and in the public view.*

we have had adequately demonstrated to us. The language of

this covenant by bitter experience we have now learned is perfectly plain; in the interpretation it has been distorted and misunderstood. The senior Senator from Idaho has been insistent in his advocacy of peace point No. 1. For it, and for the other 13 peace points, we have been told that our soldiers gladly and cheerfully fought and died in France, but the senior Senator from Idaho, and others who thought as he did. with a singular obtuseness, have insisted "open covenants of peace, openly arrived at" mean what the words import. We should have known when we distinctly announced this particular point, for which our soldiers suffered and bled and died, we meant it in a Pickwickian sense only, and spoke the usual language of diplomacy. Anyone not affected with a mental strabismus ought readily to understand that when we say "open covenants of peace, openly arrived at," we mean "secret covenants of peace, secretly arrived at," and, after their arrival, doled out in homeopathic doses to an ignorant and ill-informed populace.

The second peace point was—

Absolute freedom of navigation upon the seas, outside territorial waters, alike in peace and in war, except as the seas may be closed in whole or in part by international action for the enforcement of international covenants.

And is quite as plain to him who reads with intelligence as was peace point No. 1.

But we need not further discuss peace point No. 2. It has been lost for a considerable period now in historical mystery, and rests in oblivion with the unrealized dreams of the Akound of Swat and the Maharajah of Ruritania. Peace points three, for the removal of economic barrier, four for the reduction of national armaments, and five for the impartial adjustment of all colonial claims, based upon a strict observance of the interests of the populations concerned with the equitable claims of the government whose title is to be determined, apparently, have been submerged in weightier questions and no longer agitate the overburdened minds of statesmen. The sixth peace point, regarding Russia, I will mention hereafter. All of the other peace points, with the exception of the league of nations, deal with territorial adjustments, as does indeed No. 5. Now, the difficulty of the President in interpreting his peace points as to territorial acquisitions is that long ago England, France, and Italy reached their conclusions, and the President is up against the contracts, signed, sealed, delivered, and in the pockets of the allies.

The secret treaties which were executed before our entrance into the war were quite definite in form, and apportioned territories to the three great belligerents, and also to Russia. These secret treaties, with the elimination of Russia, I feel sure, are deemed effective and binding; nor do I doubt that substantially the terri-tory they embrace will be divided in accordance with their provisions. If the President can break through the traditions of diplomacy; if he can revive and have adopted his peace points apparently discarded with the opening of the peace conference; if he can effect a disregard of the secret treaties entered into for territorial dispositions and acqui- sitions by the three great powers; and if he can compel the altruistic and idealistic peace of which he has so often spoken his will be the greatest achievement of any statesman of any time. My hopes and my prayers are with him in his extraordinary task.

But Russia was the subject of the sixth definite peace point enunciated by the President. From the time of the Russian revo- lution until long after the control of Russia by the soviets and the cruel administration of Lenine and Trotsky the President often spoke to Russia and the Russian people, and always in gentle, kindly, and friendly fashion. His policy, which, by our acceptance, became the policy of this Nation, was distinctly enunciated in the memorable address of January, 1917. Although this policy, appar- ently, has been forgotten in our dealings with Russia, whether it will ultimately be carried out may, perhaps, be determined by the fate of the other definite terms of peace prescribed by the Presi- dent. The Kerensky government fell November 7, 1917, and the soviets, with Lenine and Trotsky in command, came into power. On the 4th day of December we were asked to declare war against Austria, and, in the course of the address demanding a declaration of war against Austria, the President adopted the formula of the Bolsheviki of Russia, the formula under which they were making peace with Germany. He said:

> *You catch, with me, the voices of humanity that are in the air. They grow daily more audible, more articulate, more persuasive, and they come from the hearts of men everywhere. They insist that the war shall not end in vindictive action of any kind; that no nation or people shall be robbed or punished because the irresponsible rulers of a single country have themselves done deep and abominable wrong. It is this thought that has been expressed in the formula "No annexations, no contributions, no punitive indemnities." Just because this crude formula expresses the instinctive judgment as to right of*

*plain men everywhere it has been made diligent use of by the masters
of German intrigue to lead the people of Russia astray and the
people of every other country their agents could reach, in order that
a premature peace might be brought about before autocracy has been
taught its final and convincing lesson and the people of the world put
in control of their own destinies.*

At this time the Russian armies had practically disintegrated
and the Russian Bolshevik! were negotiating with Germany. On
January 8, while Russia was in collapse, and it was obvious that
those in control of her government were about to make a shameful
peace, the President, with full knowledge of the situation, presumably, had this to say:

*There is, moreover, a voice calling for these definitions of principle
and of purpose which is, it seems to me, more thrilling and more
compelling than any of the many moving voices with which the
troubled air of the world Is filled. It is the voice of the Russian people.
They are prostrate and all but helpless, it would seem, before the grim
power of Germany, which has hitherto known no relenting and no
pity. Their power, apparently, is shattered. And yet their soul is not
subservient. They will not yield either in principle or in action. Their
conception of what is right, of what it is humane and honorable for
them to accept, has been stated with a frankness, a largeness of view,
a generosity of spirit, and a universal human sympathy which must
challenge the admiration of every friend of mankind; and they have
refusal to compound their ideals or desert others that they themselves
may be safe. They call to us to say what it is that we desire, in what,
if in anything, our purpose and our spirit differ from theirs; and I
believe that the people of the United States would wish me to respond
with utter simplicity and frankness. Whether their present leaders
believe it or not, it is our heartfelt desire and hope that some way may
be opened whereby we may be privileged to assist the people of Russia
to attain their utmost hope of liberty and ordered peace.*

This was said after two months of government by the bolsheviki and after they had publicly proclaimed to the world their mode
of government and what it was their purpose to do.

And, lest ye forget, I quote again the sixth of the celebrated
peace points, uttered January 12, 1918:

*The evacuation of all Russian territory and such a settlement of all
questions affecting Russia as will secure the best and freest cooperation of
the other nations of the world in obtaining for her an unhampered and*

unembarrassed opportunity for the Independent determination of her
own political development and national policy and assure her of a sincere
welcome into the society of free nations under institutions of her own
choosing; and, more than a. welcome assistance also of every kind that
she may need and may herself desire. The treatment accorded Russia by
her sister nations in the months to come will be the acid test of their good
will, of their comprehension of her needs as distinguished from their own
interests, and of their intelligent and unselfish sympathy.

On February 11, 1918, when the proceedings of the peace conference between the Germans and the bolsheviki were fully known
and understood, the President thus spoke to us:

Peoples are not to be handed about from one sovereignty to
another by an international conference or an understanding between rivals and antagonists. National aspirations must be respect-
ed. Peoples may now be dominated and governed only by their own
consent. "Self-determination" is not a mere phrase. It is an imperative principle of action which statesmen will henceforth ignore at
their peril.

Self-determination has fallen by the wayside with, "open covenants of peace, openly arrived at." Or, perhaps, self-determination
may have been used with the same delicate shades of meaning
which have distinguished the use of "open covenants." All through
our idealistic statements in this war runs the theme of self-determination and the rights of people for themselves to determine their
own forms of government. It was the underlying principle which
stimulated our judgments and aroused our enthusiasm. It is not the
fault of those who coined the phrase, and those who have used it,
that we may, perhaps, have misunderstood it, and that self-determination really means determination by ourselves of the kind of
government others should have and then impressing that kind of
government upon an unwilling and a rebellious people. With this
possible explanation, we may see with greater clarity the reason
for our activities in Russia. During February, the peace terms at
Brest-Litovsk had been concluded and there remained but the
ratification of those peace terms by the all-Russian congress of the
Soviets.

I now deal with the historical events of the fateful month succeeding the President's address. So far as I am aware, what I now
present has not been published in this country. I present it because
it is the truth, and because upon us rests a heavy responsibility for

what has since transpired in Russia. I will not be put in the attitude
of defending in any degree the soviet power or Lenine and Trotsky.
Their fantastic theories no sane man, in my opinion, can indorse.
But our dealing with Russia and the dealings of the allies with
Russia have been not only an exhibition of the crassest stupidity,
but have contributed to the awful tragedy there. Early in March
the soviet government officially presented to the allies certain
questions, favorable answers to which every man then in Russia
agreed would prevent a ratification of the Brest-Litovsk treaty by
the all-Russian Soviet congress, and then there would have been
a renewal of the war by Russia against Germany. A translation of
the official document thus transmitted to the allies from the soviet
government is as follows:

> *In case (a) the all-Russian congress of the Soviets will refuse to
> ratify the peace treaty with Germany, or (b) if the German Govern-
> ment, breaking the peace treaty, Will renew the offensive in order
> to continue its robbers' raid, or (c) if the Soviet government will be
> forced by the actions of Germany to renounce the peace treaty—before
> or after its ratification—and to renew hostilities—*

> *In all these cases it is very important for the military and polit-
> ical plans of the Soviet power for replies to be given to the following
> questions:*

> *1. Can the Soviet government rely on the support of the United
> States of North America, Great Britain, and France in its struggle
> against Germany?*

> *2. What kind of support could be furnished in the nearest future,
> and on what conditions—military equipment, transportation sup-
> plies, living necessities?*

> *3. What kind of support would be furnished particularly and
> especially by the United States?*

> *Should Japan—in consequence of an open or tacit understand-
> ing with Germany or without such an understanding—attempt to
> seize Vladivostok and the Eastern-Siberian Railway, which would
> threaten to cut off Russia from the Pacific Ocean and would greatly
> impede the concentration of Soviet troops toward the East about the
> Urals—in such case what steps would be taken by the other allies,
> particularly and especially by the United States, to prevent a Japanese
> landing on our Far East, and to insure uninterrupted communica-
> tions with Russia through the Siberian route?*

In the opinion of the Government of the United States, to what extent—under the above-mentioned circumstances—would aid be assured from Great Britain through Murmansk and Archangel? What steps could the Government of Great Britain undertake in order to assure this aid and thereby to undermine the foundation of the rumors of the hostile plans against Russia on the part of Great Britain in the nearest future?

All these questions are conditioned with the self-understood assumption that the internal and foreign policies of the Soviet government will continue to be directed in accord with the principles of international socialism and that the Soviet government retains its complete independence of all nonsocialist governments.

The men who were then in Russia familiar with the situation, including the representative of England and the representative of the United States, advised their Governments to respond favorably. I have copies of the telegrams which were sent at that time to different Governments. But the communication was received, apparently, in indignant and contemptuous silence. I can hear the voices in this Chamber saying that this was right, that there should be no compromise with crime; that the allies were correct in their position in utterly refusing to deal, or to talk, or to communicate with this awful soviet government. That it would have been in derogation of our dignity and would have been an affront to civilization to have responded, or to have united with this Soviet government in fighting Germany. I may think that you are entirely right in what I thus see passing through your minds. I may Wholly agree with you that the civilized nations of the world could not without defilement touch this awful thing in Russia. And if this had been the attitude of our Government and other Governments, we might pass the incident as one of those unavoidable historical occurrences where idealism and civilization's champions must of necessity hold aloof from civilization's assailants. But during this period what was the Government of the United States doing? It could not, you say, have any dealings with these men who had set at defiance, in the language of one of the Senators upon this floor, the laws of God and man alike. In its virtuous indignation it could permit no communication of any sort, you assert, with cruelty, rapine, murder, and anarchy. But just before the Brest-Litovsk treaty was to be presented to the all-Russian Soviet congress the United States Government went to the city of New York, interviewed the socialists and extremists, and even anarchists, who preached law-

lessness and destruction, and begged them for telegrams to be sent
to Lenine and to Trotsky, and to the Bolshevik government asking
that the Bolsheviki continue the war. I read the headlines of one of
the papers at that time:

> *United States lets "Reds" send messages to Bolsheviki. State
> Department cables their appeals to Russians to stand firm.*

And in the body of the article appear these things:

> *The United States Government yesterday aided radical socialist
> and pacificist organizations in New York in sending to the Russian
> revolutionists cable messages urging them to resist the German
> invasion.*

> *The messages were sent not only with the approval of the
> Government, but through the Government's agencies and at the
> Government's expense. * * * These messages were gathered by a person
> designated by the authorities and were sent to Washington to be
> forwarded through the State Department to Petrograd.*

The socialists and extreme radicals in New York City were
importuned for messages, and most of them responded. All of the
messages were sent by the United States Government to the Bolsheviki at Petrograd. This, of course, is common knowledge, and has
been common knowledge since its occurrence. I am not Criticizing
it. I am stating now something of the truth which we have carefully
concealed during these months. You may say to me, of course, that
this was a very different thing from communicating with the soviet
government itself, and thus, in a measure, recognizing it. You may
say that the exigency and the emergency justified the Government
in using every means in its power to continue Russia's resistance
to Germany. I may agree with you. But, if the emergency and the
crisis justified the Government in using the anarchists of New York
to communicate with the anarchists of Petrograd, it would have
justified the United States Government answering the communication of the soviet government, without recognition of that government at all, and in encouraging it, by its answer, to renew the war
with Germany. A very different proposition, you insist, between the
Government itself communicating with the soviet government, and
the anarchists of New York communicating with their comrades in
Petrograd. Perhaps so.

But, beyond this, what did we do? With an authoritative and
an official communication before us, which we were advised meant

if we replied favorably that the soviet government would make war upon Germany, we refused to respond at all. And yet just before the meeting of the Soviet congress which was fixed for the week of Tuesday, March 12, the President of the United States himself cabled the Soviet congress.

Oh, we could not touch this awful thing in Russia! We could not touch it, with the document before us by which war might have been renewed, and under which the Brest-Litovsk treaty might have been repudiated in the all-Russian Soviet congress in the week of March 12, 1918. We could not be contaminated by touching such a thing, even for so good an end. But on March 12 the Official Bulletin of the United States Government tells us this:

> *President assures Russia that United States will aid in restoring its sovereignty. Following message from the President of the United States to the people of Russia through the Soviet congress has been telegraphed to the American consul general at Moscow for delivery:*

> *"May I not take advantage of the meeting of the Congress of the Soviets to express the sincere sympathy which the people of the United States feel for the Russian people at this moment when the German power has been thrust in to interrupt and turn back the whole struggle for freedom and substitute the wishes of Germany for the purposes of the people of Russia.*

> *"Although the Government of the United States is unhappily not now In a position to render the direct and effective aid it would wish to render, I beg to assure the people of Russia through the Congress that it will avail itself of every, opportunity to secure for Russia once more complete sovereignty and independence in her own affairs, and full restoration to her great role in the life of Europe and the modern world.*

> *"The whole heart of the people of the United States is with the people of Russia in the attempt to free themselves forever from autocratic government and become the masters of their own life."*
> *(Woodrow Wilson.)*

During that week the meeting of the Soviet congress was postponed for several days in the hope that from the allied Governments would come a response to the proposition submitted, upon which a repudiation of the Brest-Litovsk treaty could be obtained from the congress. I do not care to speculate upon what might have transpired if the appeals of the Englishmen and Americans who were then in Petrograd to their Governments had been successful. I am relating to you only the facts. During this critical time in Russia the Germans were advancing through Finland. Marching and fighting with them were the Finnish White Guard. The army of the Bolsheviki were called the "Red Guard"; that of their opponents in Finland, the "White Guard."

We have been very tender of Finland, and a part of the hundred million dollars recently appropriated to feed Europe is to be devoted to Finland. The historical fact, however, is that the only soldiers who fought the advance of the Germans were the Red Guard of the Bolsheviki. They whipped the White Guard, representing the so-called government of Finland, and they were forced back only when that government and the White Guard called the Germans to their aid. It is a historical fact, as well, that wherever the Red Guard fought those of their own country opposed to their revolution they won. They were whipped only when their opponents called in German troops. This was so not only in Finland but in Ukrainia, too. During this period repeatedly it was Stated that the allies would intervene.

Intervention was never predicated upon the score of guarding supplies until the disingenuous August announcement of our Government. Intervention was suggested early in the year, and throughout the early months of 1918 it was a matter of common rumor and frequent discussion. During March the men who were most familiar with the Russian situation on the ground vigorously protested against this intervention, and no one protested more vigorously then, and at subsequent times, too, than did the ambassador of the United States Government. At Murmansk, early in March, because of the advancing Germans, the situation was acute, and then what happened? A mutual arrangement was made between the French and British, and the Russian Soviets for the defense of the district of Murmansk. And during this month of March, nothwithstanding the horrible doctrine of the Bolsheviki, notwithstanding with an iron hand they were suppressing opposition, there was cooperation between them and the representatives of the Allied Governments in Petrograd. I learned, incidentally, that Maj. Thomas D. Thacher, of the law firm of Simpson, Thacher & Bartlett, New York, who was a member of the Red Cross mission to Russia, had been at Murmansk, and had seen the cooperation there existing between the French and British and the Bolsheviki, and on the 20th day of January I wired him. asking him succinctly to advise me of the facts. I received in response this telegram from him:

One company British marines, about 130, landed Murmansk March 5 or 6, pursuant invitation Murmansk Soviet, acting under telegraphic instructions Trotsky to accept all necessary aid from allies. Before landing supreme military command Alexandrovsk and entire Murman railway granted by Soviets to committee, composed of Bolsheviki sailors, French military officer, and British military officer. This committee subject, however, complete control Murmansk Soviet. From this time until my departure March 22, French, British, and American military authorities in daily cooperation and complete harmony with local Soviet. Seventeen guns fired by British ship Glory as salute to Russian flag. Only flag visible red flag. Salute answered same number of guns, Russian cruiser Chesmar. Red Cross received effective and invaluable assistance from Soviet authorities at Murmansk as well as everywhere else in protection and transportation of supplies. Five hundred Czecho-slovak troops in Murmansk, assisted by Soviet authorities, returned to France. Sailed on my ship. I trust that this will give you the necessary information. . I neglected to add a statement of the fact that the Soviet authorities cooperated with the allied authorities in the protection of supplies at Kandalaxia, as well as at Murmansk. The supplies at Kandalaxia were, of course, more exposed to the attack of Finnish while guards, who were then cooperating with the Germans.

I received yesterday a letter from Col. Raymond Robins concerning the closing paragraph in the story of George Creel recently published. That closing paragraph of Mr. Creel is as follows:

Russia was a hard fight to Lose. It seemed for a while that we should surely win, and to this very day I believe that the people were with us then and are with us now. But the political control of Lenine and Trotsky abolished the freedom of the press and turned the power of the Government against us. Our men were the last to leave Petrograd and Moscow. Mr. Sisson stayed until the ultimate danger to secure the documents that proved Lenine and Trotsky to be German agents and then slipped out through the far reaches of Finland. Mr. Bullard and his force transferred to Vladivostok, from which point he began working back into European Russia with the allies.

One real result was achieved. Intensive work was carried on in the prison camps of Russia, and thousands of Czecho-Slovaks and Jugo- Slovaks learned of the purposes and power of America, receiving thereby the courage and inspiration that had its climax in the march of the Czecho-slovak legion from the Ukraine to Vladivostok.

Col. Robins writes me:

The inclosed clipping from the last page of George Creel's eight pages of self-laudation in February's Everybody's Magazine, suggests that Mr. Creel is unable to relate facts. The facts are:

The President's speeches were printed on the Bolsheviki government presses by special permission of that government. They were distributed under Government frank and posted on the dead walls of Petrograd by the Bolsheviki posting service.

The money for this work was drawn from the Petrograd State Bank by special O. K. of the Bolsheviki government. The President's speech of January 8 was sent over the private wire of the Bolsheviki Government to Brest-Litovsk by order of Lenine, and I secured the order for its transmission in the presence of Mr. Sisson.

Mr. Sisson fled from Petrograd on the 4th of March, 1918, shouting that the Germans would take the city within a few days in collusion with the Bolsheviki. The American Red Cross was feeding starving children and evacuating war supplies from Petrograd under Bolsheviki protection in quantity until the 1st of May, 1918.

The allied military missions were helping to train the Bolsheviki Red Army on the 1st of April, and the American ambassador was seeking. with the consent of the allied embassies, the cooperation of the American Railway Mission with the Bolsheviki government, weeks after Mr. Sisson had fled in terror from Petrograd.

Mr. Bullard and all the American members of the committee on public information in Russia fled from Moscow on the 5th of May, reaching Archangel and suffering from the worst case of "buck fever" in my observation. They telegraphed the American ambassador to got them the permission of the British high commissioner, still in Petrograd, to get on the English ice breaker, which permission I secured for them, and they were on this ice breaker dressed in English uniforms for several weeks, while the American Red Cross was doing business as usual in Moscow, and English, French, and Japanese were still working for their Governments under the protection of the Bolsheviki government.

The American Red Cross mission did not leave Moscow until the 5th of October and Petrograd on the 16th of October in the full protection of the Bolsheviki government, some seven months after Mr. Sisson had fled in terror with his panic from reading some propaganda pamphlets and papers which were the amusement and contempt of honest and informed men.

When last I addressed the Senate upon this subject I asked
certain questions of the Government. It is unfortunate that there is
no method by which our Government may be interrogated and no
means, except that of a majority vote here, by which information
may be obtained for our people. I wish it were possible so to amend
our system that up-on this floor the appropriate official could be
asked legitimate questions and compelled to respond. No answer
has been made, of course, to the questions that I asked, because the
implication in every question is known by the department to be
true. Men are in this country to-day who can establish every single
fact suggested, and if this body, or the Foreign Relations Committee, really desire information, if they wish to tell the mothers and
the fathers and the wives of the men who are freezing in Russia to-
day just what the facts are, they have at their disposal the evidence,
and it can be brought before them upon the briefest notice.

Russia, Mr. President, is a marvelous country. It contains
one-sixth of the earth's surface, with fertility of soil and wealth in
mineral resources surpassing those of any other part of the earth.
This Soviet government to which our President for us spoke so
kindly, begged us for economic aid and wished to make us the
most-favored Nation. We, in the rigidity of our virtue, though asking the aid of the anarchists in our midst, with the Bolsheviki over
there, and though publicly proclaiming our friendship and our love,
would not accept the proffer. Weak and vacillating, stupid, and ignorant has been our policy with Russia. We solemnly promised we
would not intervene, and then, prating of our love for the Russian
people, we did intervene. Prating about guarding stores at Archangel, we advanced from 100 to 300 miles from that port, took and
burned little Russian towns, and upset little Soviet governments.
In the name of protecting military supplies, which were offered to
us again, and again, and again, and which we could have had for
the asking, we shot down Russian peasants, and our boys are shot
down by them. The Senator from Nebraska insists our only purpose
in landing at Archangel was to protect the stores. Our only advance
beyond Archangel was to prepare military bases. He is wrong. We
were marching down from Archangel—and the facts will demonstrate it—that we might make conjunction with the Omsk Government and might perfect the ring of steel which we had thrown
around interior Russia, and which was starving innocent women
and children. The Senator from Virginia gravely speaks of the
German menace. What German menace since November 11? Are

our people children to be lulled into repose by such stuff as this? The very learned and logical Senator from Colorado tells us that we are not making war upon Russia; that Russia is making war upon us. Apparently his argument seems to be that if the Russians had not resisted when we advanced into their territory there would have been no conflict and no killing. What a strange and fantastic doctrine is this! If an army landed in New York, marched to Buffalo, and the people in central New York resisted and fought them, by that fact, then, New York was making war upon the invading army and the invading army was innocent of wrong. The French are under no illusions in this matter.

They are for intervention, and they believe they are intervening upon a small scale, too small, as they put it, now. They make no pretense that they wish supplies guarded. They wish Russians killed and another government set up. What hypocrisy upon our part to say to our people, and to the Russians, in our pronunciamento last year when we commenced our intervention, that we contemplated "no interference with the political sovereignty of Russia, no intervention in her internal affairs, not even in the local affairs of the limited areas which her military force may be obliged to occupy, and no impairment of her territorial integrity, either now or hereafter." No sooner had we landed at Archangel than we shot the Soviet government there existing out of town and set up a government of our own. No sooner did we go into the interior than everywhere we found a local soviet we shot it to death and set up our own mode of government. Then we tell our people that we intend no interference with the internal or local affairs of Russia!

What a commentary it is upon the power of this body and upon the power of our people that the State Department and the (Government can not and will not answer concerning our troops in Russia. How the iron must enter the souls of those who have relatives there; of the mothers and fathers and the wives of men who were drafted to light Germany, and then, when the war with Germany was ended, were forced to fight a war with Russia. Our Government can not answer concerning our troops, although the great preponderance of the forces at Archangel are American. Of necessity it must respond in indefinite and general terms. These troops are now under English command, and the Americans who are in the vicinity of Archangel, like good soldiers, are obeying their English commanders. I venture the assertion that in Washington, in the departments here, they know no more about what military action is contemplated or what our troops are about to do than the veriest novice upon the floor of the Senate. The American troops are under English command. Perhaps justly so, and I have no doubt well commanded; but it is because of that fact that no answer can be made as to the position of these troops or their military activities. An occasional belated story is given to us. the last of which we read on January 25, as follows:

> "We have reports from Archangel," said Gen. March, "which were received here on January 24, and have been decoded. We had at Shenkursk a force which, at last reports, consisted of a detachment of British, two American companies, and two Russian companies. Manifestly this force has had out in that vicinity small patrols at times. The towns mentioned in the report are so small that we can not find them on our largest maps. The force at Shenkursk was attacked on three sides, and the report which was received at Archangel that day stated that they were forced to evacuate.

> "The troops at a place called Ust-Pedenga, which I can not find on the map, were also compelled to evacuate under attacks by strong Bolshevist forces. Our troops took up a position midway between Shenkursk and Ust-Pedenga. Under attacks of 1,000 of enemy troops we retreated from Tania to a point 10 miles away therefrom. Tania is 18 miles west of Shenkursk.

> "Under an attack from 200 of the enemy forces we retreated from Kodema, which is 25 miles from Shenkursk, to a point within 13 miles of that town. The enemy has strong patrols from Shegovari to Tania. Shegovari is 20 miles north of Tania, and to the right of Shenkursk.

> "The Americans lost 10 enlisted men killed in action, 17 wounded, and 11 missing in the retirement. Near Ust-Pedenga, and later at Shegovari an enemy attack on the west was repulsed."

> "Do we plan," was asked, "to reinforce the American unit in Russia in view of the fact that it seems to be retiring?"

> "Shenkursk," replied Gen. March, "as scaled on the map, is apparently 100 miles from Archangel, and the allied forces representing four Governments, and the Russians, five Governments, have up there a force large enough to reinforce those men or have them fall back on them and hold the situation.

Gen. March was asked whether any recommendation had been received concerning the withdrawal of the American forces from the Archangel front.

The force now at Archangel—

He said—

*was put in there by the allied Governments on the recommen-
dation of Marshal Foch, and the military handling of that unit was
thereupon turned over to the supreme commander; and whatever is
done concerning reinforcing the unit will be done by him. The allies
agreed upon a British commander in chief, and at the last reports he
was in the front lines Inspecting his troops, and I assume that the military part of it is in hand.*

From this, apparently, all the roseate stories of the campaign
in Russia may be discounted. Our brave boys had to retreat many
miles through the ice and the snow and the rigors of an Arctic
winter. They had to fight during this period in weather that few are
accustomed to and none understand.

And what has been the result of it all?" It is true we have a ring
of steel that prevents food getting into Russia. It is true that we
are starving women and children to-day in Russia, and that hands
are lifted in supplication to God and in cursing this country for its
activities. But is it not obvious to you that when we compel people
in Russia to starve, who will starve? It will not be first the horrible
Bolsheviki. It will not be your Lenines and your Trotskys. It will
be the intellectuals and the bourgeois. It will be those whom we
are pleased to term the very best people in Russia. As you contract
this ring now, and as you prevent the natural flow of food from the
granary of Russia, and you blockade her ports with gunboats, you
are first starving the very people you do not wish to starve, and
you are starving those who ought not to starve. If it had not been
for this criminal policy of intervention, and this ring of steel that
prevented food getting into internal Russia, Lenine and Trotsky, in
my opinion, long ago would have fallen and the Bolsheviki would
have been at an end. But by this foreign interference every base
passion has been appealed to, every prejudice aroused, and even patriotism invoked. And the very act of intervention has enabled this
grotesque government to last far beyond its allotted time, and to
exercise its despotic sway in the name of public safety. How much
have we contributed to the terror, to the rapine, plunder, slaughter,
and massacre? There is a heavy reckoning some day for those who
have been responsible for this wicked and this useless course in
Russia. And the heaviest responsibility, the wrong which can never
be atoned, is the shedding of American blood in Russia. It is to this

phase I desire to arouse the Congress and to which, if I had the power and my voice would carry, I would arouse the people of the Nation. It is of American boys and American blood I am thinking. I would not give one American life in Russia for all the bolsheviki spawned by centuries of tyranny and mad with the lust of a ruthless ephemeral power.

What I ask by this resolution is that our Government, which shrouds itself in mystery and which will not tell us or the people its intentions or its policies, may know our opinion that our troops should be withdrawn from Russia. I do not care how you view the situation, whether you favor armed intervention or whether you do not. If you favor armed intervention it is obvious that the scale upon which it has been undertaken is too small to accomplish lasting results. It has become painfully clear in the last few days that by the present intervention we merely hazard the lives of our men. It is equally clear that the people of England, and our own people, will not tolerate intervention upon a larger scale. Therefore, even if you favor intervention you should, for the protection of the lives of our soldiers, insist that those there, few in number and their position courting disaster, be immediately withdrawn. If you favor intervention, why do you not upon this floor, by resolution or otherwise, say so? If you believe in war with Russia, why not introduce an appropriate resolution and permit Congress to vote upon it in accordance with the Constitution, and permit our people to discuss it? Upon what theory can you justify war, without affirmative action by Congress? And that we are in an actual state of war at present in Russia the recent ominous news from there demonstrates only too plainly.

What a sorry spectacle we present! The distinguished chairman of the Foreign Relations Committee endeavors to make explanations of our policy with Russia, but hastens to assert that he does not speak officially or authoritatively. The chairman of the Committee on Naval Affairs gravely gives his view s but prefaces them with the statement that he speaks neither officially nor authoritatively. Both of these gentlemen say we are simply protecting supplies. I reiterate that England is under no illusion respecting our purpose in Russia, and that France frankly proclaims it. The court organ, the New' York World, says we are in Russia for the purpose of maintaining order. The various Russians who have showered us with pamphlets of late understand the situation full well. The most recent of the propaganda which has come to me is from a certain

Col. Vladimir I. Lebedeff, who says:

> *Just at this time, the allied armies being at Vologda, the allies advised the officers' organizations to revolt simultaneously against the bolsheviki in Ribinsk, Vladimir, Yaroslavl, and Murom, so as to encircle Moscow as with an Iron ring.*

In the declaration published by the State Department of our intervention it was stated "whether from Vladivostok or from Murmansk and Archangel, the only present object for which American troops will be employed will be to guard military stores which may subsequently be needed by Russian forces and to render such aid as may be acceptable to the Russians in the organization of their own self-defense." The charming naivete of this must have appealed to the Bolsheviki. We were landing troops for the sole purpose of guarding military stores which subsequently we would deliver to the Russians themselves. Of course, the intention of this utterance was to fool the people of the United States into believing American bayonets were necessary to protect Archangel stores from the Germans. It is never for an instant indicated nor is the language susceptible of any such meaning that these stores were to be protected from any kind of Russians who were not acting with the Germans. Our advance constantly into the interior, of course, makes it obvious that the guarding of supplies for Russians themselves was the veriest kind of pretense. In all of the months prior to our intervention daily the subject of intervention was hotly discussed.

In all these discussions there never was a question about protecting supplies except in the instance where the soviet government, in conjunction with the English and the French, at the very moment of the German advance in February and March, protected supplies at Murmansk. The distinguished chairman of the Foreign Relations Committee scoffs at the idea that any war exists in Russia now, or that we are there participating in war with the Russians. I take it what he means is that it is just a little war, and because it is such a little war, it ought to be disregarded. It is a little war. It involves the lives of something over 5,000 American boys only. It touches the hearts of perhaps three times that number of anxious relatives in

the United States. It is true, perhaps, that it is such a little war it involves only a few hundred lives, a few hundred maimed, a few hundred American graves on Russian soil, but to me, sir, one of those lives in Russia and one of these hearts bowed in anguish in America are more precious than all of the pretense of diplomacy or the protection of any government from the just wrath of a righteous public opinion.

Opinion upon this subject has crystallized in England. You may have observed recently a delegation of soldiers called upon Lloyd-George and expressed their fear that they might be sent to Russia. He assured them they would not be. I have been interested in following the discussion which has been going on in England, and a portion of it I wish to bring to your attention. Indeed, as evidentiary matter, I have selected two newspapers, one British the other American, because of their high standing and the accuracy of their statements. I have taken the Manchester Guardian, one of the few great independent newspapers in the world. It certainly will not be accused of Bolshevism or sympathy with men like Lenine and Trotsky. I have taken as well from the sedate precincts of the Old Bay State the Springfield Republican with its recent reviews of the Russian situation, and, without at length reading either, I desire to refer to a few excerpts.

Perhaps, Mr. President, I would serve no good purpose in reading these excerpts, and, if the privilege will be accorded to me, I should like to put into the **RECORD**, as a part of my remarks, various articles from the *Manchester Guardian*, dated December 6, December 19, December 24, and December 27, 1918, and two articles from the Springfield Republican, one of them dated January 5 and the other dated January 12, 1919.

The PRESIDING OFFICER (Mr. GAY in the chair). Without objection, it will be so ordered.

The articles referred to are as follows:

[From the *Manchester Guardian*, Dec. 6, 191S.]
THE RUSSIAN SCANDAL.

Our present relations with Russia are about as indefensible as can be imagined. The foreign office, or those who control the policy of the foreign office, can hardly be ignorant of that—how should they be?—but, so far as appears, are preparing to cover one error with a greater error and to make bad worse. In the midst of the preoccupation of a great war the little war with Russia has received comparatively little attention, and it is hardly realized that, though the great war is over, the little war goes on, and, moreover, that if it is not stopped now or soon it is likely to become a very much larger war and a more and more intolerable and indefensible one, so intolerable and indefensible, indeed, that it is capable of producing grave reactions here, extending to the overthrow of a government. We originally embarked on this Russian adventure under wholly different circumstances, and for reasons, so far as reasons were given, which have no present application whatever. After the intervention by Germany in the civil war between the Reds and Whites in Finland and her virtual occupation of the country there was a real, if somewhat remote, danger that she might strike through Finland at the narrow strip of Russian territory which divides the north of Finland from the Arctic Ocean and establish for herself a naval station on the Murmansk coast, giving her access to the Arctic and a new outlet for her submarines. At a moment when the submarine war was at its height and Russia lay helpless and subservient there was reason, if not very urgent reason, since the district was extremely inaccessible and a long line of railway would have had to be built, for guarding against this peril. That was the extent of the danger and the extent of the need for our occupation. Its extension to Archangel, which is not ice free, and to a large stretch of country inland had no such justification. The occupation of Vladivostok, Russia's ice-free port in the Pacific, 5,000 miles away, followed by the advance with the cooperation of Japan westward into Siberia was, as a military measure, equally unjustified. Both these extensions of the original intervention were defended on quite other grounds. It was said that Russia had become the mere tool of Germany and that it was necessary, first, to prevent the further extension of Germany's influence and her Increasing exploitation of Russian resources, and, secondly, to "reconstitute the eastern front." In this connection the happy discovery was made of scattered bands of Czecho-Slovak prisoners who, it was urged, must, in the first place, be rescued, and, in the second place, utilized in this process of reconstituting the eastern front.

So matters stood at the time of the collapse of Germany and the conclusion of the armistice. Obviously, every reason hitherto alleged, whether for the original occupation of the Murmansk coast or for the subsequent expedition to Archangel and Siberia had now disappeared. They were all in the nature of defenses against the attack of Germany, and there was no longer any attack or possibility of attack from Germany. Germany as a military power was dead; but were the defensive measures, the counter-attacks, dead also? Not at all; they continue in full force. There is quite a prospect that they may be largely extended. For the moment, of course, there is a pause. Winter is no respector of persons or of policies. Very soon Archangel will be frozen up and our army of occupation there will be frozen up also. In the east the Japanese have steadily and very sensibly refused to advance a mile further. They have reached Lake Baikal, and beyond Lake Baikal they decline to go. So if we desire to extend our operations in this direction we shall have to do it ourselves, for America will certainly not assist us and will prudently follow the Japanese example. But there are other possible fields of operation. We have recently obtained access to the Black Sea. We are therefore now in a position to repeat in the extreme south of Russia our performances in the extreme north, and, as a matter of fact, it is credibly reported that the war office is now engaged in making a survey of the country. The Ukraine, under German and Austrian control, has become the refuge or dumping ground of a whole collection of Russian reactionaries of various sorts and sizes, and the same is true to a less extent of the Don country and other districts to the east. It would be easy to play into the hands of these gentry, as we have played into the bands of others of the same description in eastern Siberia, where a purely reactionary party has now dismissed the local popular (not Bolshevik) government and established a military government of its own.

But what conceivable justification, it may be asked, is there for any such proceedings? And how is it possible that any British Government should embark on so wanton and criminal an interference in the affairs of another nation? Such questions may, indeed, well be asked, but it is doubtful if they will receive any presentable answer; for the fact is that the real, though unavowed, reason for our previous interference is entirely different from the avowed reasons, and it is a little awkward, now that the avowed reasons have disappeared, to produce the real one—the more so as this is not a very nice reason or one which is likely to commend itself to reasonable people in this country or to our working class. That reason, of course, is that the war against Russia has from the first really been a war against that particular form of socialistic theory known as Bolshevism.

We are no admirers of that theory. Applied to any western European country, we believe it would be disastrously subversive. Even in Russia we may doubt its permanence. But there it is; it has established itself; it has existed for more than a year. It is not weakening in power; all trustworthy information goes to show that it is gaining in power; that it has established order; that it meets with general support from some 80,000,000 people, whom it controls; that it is grappling successfully with the food problem; that it is promoting the popular arts, music, and the drama, and is preparing a great scheme of popular education—that, in fact, it is performing most of the normal functions of a government, and performing them with increasing success. These are the facts, but they do not suit the policy—the policy at least hitherto pursued—of our freedom-loving Government. The telegrams of the few British correspondents, including our own, who are still in a position to give authentic information, are ruthlessly censored or suppressed, and the Government goes on in its blind and foolish way, a way that can, if persisted in, lead only to discredit and disaster. This is the situation as we understand it. Bolshevism is to be suppressed by armed force, and in order to prepare people's minds for it and to lend it some color of justification, not only is truth as to the condition of Russia suppressed but currency is given to all kinds of wild statements and palpable exaggerations. The Bolsheviks are not angels from heaven. They have, like most revolutionaries, executed a good number of their enemies, but these executions have mostly taken place since the intervention of the allies gave encouragement to the counter-revolutionaries and made them more formidable. Mr. Litvinoff, who is an honest man and a Tolstoyan before he is a Bolshevik, puts the total number of executions since the Bolsheviks came into power at 400, half of them of ordinary criminals. That is probably an underestimate. If information were allowed to come through from other sources, we might get nearer the truth. This, then, so far as there is a policy, appears to be the policy. It has got to be changed. Perhaps the Government are already awakening to the fact, but find it difficult to get out of the mess they have themselves created. Let them take heart. It is easier now than it will be later. Every week, every month, that they stay in Russia and wage a war on Russia, which has lost every shred of avowable reason and has no justification, will make it more difficult to escape. If with the coming of spring they should see fit to resume or extend their military operations, it is well they should understand that it will not be tolerated in this country. The workers here are not going to send their sons to slaughter and be slaughtered against the workers of another country, against which we have never even declared war, and for the purpose of destroying a form of social economy with which some of them, at least, are in sympathy. If it is sought to check the progress of Bolshevism and prevent its spread to this country, that is precisely the way to defeat the object. We prefer not to consider the possible further consequences of such reckless folly.

[From the *Manchester Guardian*, Dec. 19, 1918.]

ADMIRAL KOLTCHAK AND LORD MILNER.

The campaign of Admiral Koltchak, the Siberian "dictator," against his old comrades in arms is developing. He has arrested another 27 of them, including M. Tchnernoff, one of the foremost of the anti-Bolshevik socialists, and 12 other members of the constituent assembly. Their crime is that they do not accept him as dictator and that they proclaim him a sheer reactionary, a charge which has that sharpest of all stings—truth. The admiral's campaign against the Bolsheviks, however, is not moving equally prosperously. The chief Czech generals have resigned rather than do his work for him, and the whole Czech army is threatening to abandon the front and go home. That would mean the end of Admiral Koltchak's "dictatorship" and the whole miserable Siberian adventure. The admiral, however, expects the allies to induce the Czechs to stay on as his mercenaries (paid, of course, not by him but by the allies), to replace them, if and when they depart, by allied conscripts (thus doing their bit to make democracy safe), and to provide him liberally with the sinews of war. The finance of the " dictator " is characteristic. It will be remembered that the late Czar gained much credit at the beginning of the war by suppressing vodka Vodka is the dictator's local financial stand-by. It brought in 1,000,000 rubles in August and 10,000,000 rubles in November, and its potentialities are "limitless," for there are a thousand million rubles of the stuff in stock. Thus, faithful to tradition, vodka is to be·the rock on which the "dictator's" Russia is to be builded, just as it was of the Czar's Russia. Vodka and loans—for the "dictator's" foreign friends are to do most of the paying, though we should like somebody to tell us who is calling the tune. Koltchak is asking the allies to provide 60,000,000 rubles a month to keep him going. There is a further point. Many parts of Russia are starving. How many people realize that it is partly the allies who are responsible for starving Russia? We are blockading Russia so that no raw materials or machinery can enter the country or produce leave it. We are sustaining the blockade of Russia by the revolted provinces who used to supply her with most of her food and fuel. To what end this martyrizing of the Russian people? To what end this pouring out of British blood and British money?

To these questions, at long last, a British minister, Lord Milner, attempts a reply. Lord Milner does not say the Bolsheviks are German hirelings. He is not quite so ready as his colleagues to adopt, on the strength of forged documents, a theory which no competent authority

*sincerely believes. He says we went to Russia for two chief reasons—
because the action of the Bolsheviks was assisting our enemies by
releasing hosts of German troops and bringing Roumania down, and
because we were under a moral obligation to save the Czecho slovaks.*

*We are afraid that these reasons ignore a great many facts. We
provided the Ukraine with money and arms to break away from
Russia before the Bolsheviks made peace with Germany; indeed,
our action was an important, perhaps the most important, factor in
precipitating the treaty of Brest. Again, the Bolsheviks repeatedly
expressed their willingness and their anxiety to let the Czecho-Slovaks
leave Russia, and they contend that it was the allies who inspired them
to stop in the hope of "reconstituting the eastern front." That, indeed,
was the favorite argument for the expeditions to Russia—that it
"would bring Russia back into the war, whether the Russian people like
it or not. We venture the assertion, in despite of Lord Milner, that the
military adventures of the allies in Russia did nothing to bring defeat
to Germany, but were in a military sense pure waste, to say nothing
of the political loss. But be Lord Milner's reasons for intervention as
good as they are Lad, they no longer exist. The war with Germany is
over. The Czecho-Slovaks can go home as soon as they like, or rather as
soon as we let them, for it is the allies who are keeping them in Siberia
when they want to go home. What then is our reason for maintaining
war against Russia? Lord Milner says that we have induced some
thousands of Russians—a very few thousand, in fact— to fight on
our side and we can not abandon them to "the unspeakable horrors
of Bolsheviki rule" until they can train, arm, and defend themselves.
As the Bolsheviks have a large army and the anti- Bolsheviki parties
are better at fighting one another than at fighting the Bolsheviks,
it looks as though we shall be at war with Russia for-ever, on Lord
Milner's principle. Would it not be cheaper to come to terms with the
Bolsheviks, safeguarding the lives of our Russian friends, which would
be perfectly easy? The Bolsheviks are begging for an armistice, but we
will not condescend to listen to them. Why? The true reason peeps out
at the end of Lord Milner's letter when he speaks of our duty to prevent
"barbarism" spreading all over Russia. In Russia we are fighting
neither against the Germans nor for the Czecho-Slovaks nor for the
Russian anti-Bolsheviks. We are fighting against a form of the State
and a conception of property which we dislike, and which we have good
reason to dislike, but which it is not our business to overthrow by force
of arms in another country. That is why we are in Russia.*

[From the *Manchester Guardian*, Dec. 24, 1918.]

THE OUTLOOK THIS CHRISTMAS.

The fifth Christmas since the war began is not yet the Christmas of peace. Fighting has been suspended between the original belligerents, nor is there any prospect of its being resumed. Turkey, Austria, Bulgaria, and Germany have neither the will nor the means to take up arms again before the final peace is ratified. All this is an immense boon, but we should not forget that the old war has either actually given place or is in danger of giving place to a new war. The allies are directing or sustaining several campaigns against the Moscow government, and the fragmentary States of the disrupted Russian and Austrian Empires are engaged in a conflict with one another, which is not the less ferocious because it is obscure. These, perhaps it will be said, are but small dark patches on an otherwise brilliant prospect. Nobody, however, knows what dimensions the war against Russia may take on, just as nobody knows what this war is about. Still more disturbing, nobody can foretell what vast explosive forces may be released and transmitted throughout the world by such a reckless enterprise. Old Europe was light of heart when it set out to extirpate revolution in France, and the end of it all was that old Europe collapsed in blood and fire under the blows of Napoleon. Of course, history may not repeat itself, but the statesman who stakes much on the chance that causes will not produce their probable consequences is not exactly prudent. All the miserable conflict in eastern and southeastern Europe between nation and nation has a very direct relation to events in Russia. Russia is girded round with primitive race passions, the fierce tyranny of intolerant pride and aggressive exuberance. That constitutes an immense mass of inflammable material. A cautious statesmanship would hasten to extinguish the war with Russia, before it extends and consumes great part perhaps of Europe. When we see European statesmanship, on the contrary, planning to develop, instead of ending, hostilities against Russia, we can not think the outlook too comforting.

The moral flaw thus revealed is as unfavorable to excessive optimism as the intellectual flaw. The war against Russia has the two characteristic defects of the pre-1914 system—the secrecy of its inception, its conduct, and its aims, and the conviction that force is the best of arguments and can be trusted to prevail.

[From the *Manchester Guardian*, Dec. 27, 1918.]

THE ALLIES AND RUSSIA.

The allies, according to M. Pichon, are desirous of extirpating Bolshevism in Russia, but they are not going to expand their military intervention. If the allies were united in extirpating poverty outside Russia, it would please the allied peoples and benefit them much more. Bolshevism inside Russia is the concern of the Russian people, not of the allied governments. We can not imagine anything more reckless than an adventure of this kind at a time like this. Some perception of this fact seems to have dawned upon the allied governments, perhaps with the help of President Wilson. That explains, presumably, the decision to resist the pressure of the Russian exiles anxious for an unlimited allied expedition. But the policy of a limited military commitment is a worthless compromise. No man knows where it will end, no man knows what result it will produce, no man can safely predict of it anything except that it will prolong civil war in Russia indefinitely, and delay indefinitely the ending of the starving of the Russian people by the blockade and the restoration of tolerable conditions. The allies can either have war with Russia or be at peace with Russia. There is nothing between. The allied governments must choose the one or the other. They would be wise to choose peace quickly.

[From the *Springfield Republican*, Jan. 5, 1919.]

FRENCH AND RUSSIAN SITUATION.

It now seems settled that France, at least, intends to persist in the unhappy half-way course which has worked such disaster in Russia. Last week mention was made of the reported abandonment of the great project of sending armies into Russia to overthrow the Lenine government. But apparently while the allies could not agree on this "thorough" method of dealing with bolshevism by attacking it at its supposed source, France can not make up its mind either to abandon intervention, Foreign Minister Pichon last Sunday flatly contradicted the inference of a Socialist deputy that this was what he had meant, and made the further statement that while intervention was inevitable its purpose " was not offensive for the time being but defensive. Later an offensive might be necessary to destroy Bolshevism, but such an operation must be carried out by Russian troops, of whom there were at the present time, he said, 100,000 at Odessa.

By "defensive" operations he meant, of course, not defending French territory, but preventing the spread of Bolshevism that is to-day

*preventing the extension of the Lenine government's power over those
parts of the former Russian Empire, which it does not yet Control. Yet
this professed purpose hardly covers the case of the present intervention
of the French, apparently without the cooperation of their allies in
the Ukraine. For the Republican forces, which the French seem after
some hesitation to have decided to attack are not pro-Lenine, and it is
doubtful if they can be called Bolshevist except in the extended sense in
which that term of opprobrium is coming to be applied to revolutionists
in general. It will be recalled that a fortnight ago we had word of the
sweeping progress of the Republican forces under the popular leader,
Petulra. They took Odessa and other Black Sea ports and also occupied
Kiev, the capital of the Ukraine. Apparently it was they and hot, as
an earlier muddled or censored dispatch indicated, the white army of
Gen. Denikin that overthrew the Cossack hetman, Gen. Skoropadkin,
who with German support had carried on a counter-revolution in the
Ukraine. On the contrary, Skoropadkin appears to have gone over to
the allies after the collapse of Germany, and to have acted upon their
advice in declaring for a reunion of the Ukraine under the government
which he was to set up, with the Russian governments organized by
the reactionary dictators, Gen. Denikin and Admiral Kolshak.*

*This was wholly contrary to the will of the people of the Ukraine,
but a Pole of high rank, who lately escaped from Kiev, is probably
mis-taken in attributing the popular revolt which followed to this
declaration for union. A much more probable cause is that Skoropadkin,
Deni-kin, and Kolchak all represent the propertied interests and
aim at undoing the work of the revolution, especially in regard to the
distribution of the land. In the Ukraine, which takes in most of the
famous "black earth" belt, in which the great landlords stuck tighter
to their vast estates than in northern Russia, the agrarian question is
specially acute, and this issue seems to have united the peasantry and
the lower and middle classes of the cities to a degree not found in Russia
proper. The movement headed by Petulra, in short, has all the marks
of a genuine popular movement, and the way in which it burst out
everywhere all at once shows that it had gained much headway before
the censors allowed mention of it.*

*When the new tidal wave of revolution swept down on Odessa
and the other seaports the French, who had no adequate forces on hand,
contented themselves with protecting the docks and wharves with their
naval vessels, and later landing a force were said to have regained
about a third of the city. Now we are told that a French army is to
operate from Odessa, and that another army, under French command,*

but presumably in the main Roumanian, is marching from Roumania to attack Kiev, the capital of the Ukraine.

Of the military side of these operations it would be premature to speak, because so little is known of the force involved. But politically it can at once be said that the adventure inspires serious misgivings, if only because the French foreign office, in some ways the most reactionary now left in Europe, has shown such unlimited capacity for blundering in its dealings with the revolution. It is now supporting a reactionary "unionist" government, yet last year the Ukraine had hardly declared its independence when the French recognized its Government, lent it $35,000,000 (surely the worst bet ever made), and sent a military mission. The theory was that since the Ukraine was against the Bolsheviki, and since the Bolsheviki were pacifists and pro-Germans, the Ukraine must be pro-ally and eager to go on fighting for the entente. But that military mission got to Kiev just in time to see the Ukraine break faith with the Bolsheviki and negotiate a separate treaty with the enemy. M. Pichon has a reason to feel sore about that $35,000,000, but it should be noted that the hasty advance of cash to a Government simply because it was revolting from the Bolsheviki showed a too rigid and fallacious logic. In real life things do not arrange them-selves as this or that; they may be something else, and Ukraine was the something else. It may be added in extenuation that French policy in Russia has from the beginning of the revolution been under inconceivable pressure from the half-crazed investor. For people whose hard-earned savings are menaced by a social cataclysm in a remote country the utmost sympathy may be felt, but there is no worse influence for a statesman than a panic over investments.

It is in France, for this reason, that we must look for the bitterest hostility toward the revolution, and strange as this may seem in view of the preeminence of France in revolution. Of exceptional interest, now that France has drifted into war with Russia, is the frank account given by the historian, Ernest Lavisse, when the revolution was but two months old, of the friction between the two countries:

"If we have misunderstood the Russian revolution, the revolutionaries, on their side, have not been fair to us. We looked for some expression of warm sympathy toward France, who first proclaimed the rights of man, and whose successive revolutions have contributed so much in the downfall of the old regime in Europe. But the Russian revolutionaries have greeted us with black looks. They reproach our Republic with having made too easy a pact with czarism

and with having permitted and even supported the very equivocal activities of the Russian police in Franco. On this last point they are only too well justified. But they don't appreciate in a spirit of fairness the reasons for the alliance between the French Republic and the Russian Czar. France lived under perpetual threat of a German attack. * * * Certainly this alliance was repugnant to our traditions, but sheer necessity forced it on us."

Hardly less serious than the situation in Ukraine is that in western Siberia, where a new danger is revealed in a dispatch from Carl Ackermann, who has been at Omsk, in which he states on November 24 the Czech soldiers "voted against an offensive against the Bolsheviki which had been planned and ordered." This is striking in the first place because it shows the Czechs, whose prowess and good conduct have won admiration, as having absorbed in Russia the Bolshevist idea of the referendum to the army of the orders issued by the high command. This detail gives point to what has been said of the infectiousness of Bolshevism and its influence upon all the armies that are sent to fight the Bolsheviki.

But for the Czechs there is the justification that they feel, says Mr. Ackermann, that they have been betrayed by the allies. Their feeling may be unjust, but at all events the allies exploited the accidental presence of the Czechs in Russia without having worked out a plan or even having agreed on a general policy, and the Czechs, like the Roumanians and the Serbians, have been left in the lurch. When the decision to have them stay in Russia was announced attention was called in these reviews to the danger of making so immensely important a decision as invading Russia, a matter to be determined on broad lines, turn on the mere accident that a small force of foreign soldiers was occupying a good strategic position thousands of miles into the interior of the empire. There were 60,000 Czechs getting on very well in a land of 160,000,000 people, because there was no serious hostility; they were Russia's allies leaving Russia. Why should there be hostility? Here and there German agents stirred up trouble, but it is now tolerably clear that there would have been no difficulty in getting the Czechs out of Russia, if that had been desired. Even the treacherous hostility of Trotsky seems to have been due to the threats of intervention and his belief that if the Czechs got to Vladivostok they would join the invading entente armies.

Of the part played by the United States Government in this sorry affair it is too soon to speak, but it is certain that its action was taken in the supposed interest of the Russian people. But the fear that its action would be misunderstood in Russia has been borne out by the result, and is illustrated in a telegram sent by the president of the soviets of central Siberia, which Mr. Ackermann quotes:

"From Russia to Vladivostok are moving 60 echelons of Czecho-slovak troops at the disposal of America, ostensibly on their way to France. In view of the hostile attitude of international imperialism and the threats of a foreign landing at Vladivostok the central executive committee of the Siberian soviets considers a concentration there of forces dangerous and inadmissible."

The American official shares in commitments which the allies did not in fact carry out is shown in a dispatch from Consul General Poole at Moscow to the Czechs at Samara:

"You may inform the Czecho-Slovak leaders that pending further notice the allies will be glad, from the political point of view, to have them hold their present positions. On the other hand, they should not be hampered in meeting the military exigencies of the situation. It is desirable, first of all, that they should obtain control of the Trans-Siberian Railroad, and, secondly, if this is assumed, at the same time, if possible, retain control over the territory they now dominate."

Is not this a curious dispatch to send—from Moscow? Has the Lenine government had altogether a square deal, to say nothing of the Czechs?

Their vote against another offensive fits with all that we have had from Siberia and helps to explain the hostility toward them shown by the London Times correspondent, who was specially angry at them for protecting the members of the constituent assembly from arrest by the new dictator, Kolchak. By their aid the delegates were able to escape to Ufa, in European Russia, where the all-Russian government had its capital before it went to join the Siberian government at Omsk. At Ufa, said the correspondent of the Times, they "continue to fulminate against the Omsk government," though their power for mis chief there was limited by lack of the support of the Czechs. But this week the Moscow government claims the capture of Ufa. Was it a military victory, or have the leaders there abandoned the allies because of the Kolchak coup d'etat and gone over to Lenine? If the allies have not been fair to the Czechs, have they been fair, either, to the Russian moderates who have been willing to cooperate in restoring Russia? There is much excuse for them if, after narrowly escaping arrest by a czarist military dictator, they have finally joined forces with the Bolshevlki; but why *should the allies have let them be forced into that dilemma?*

The refusal of the Czechs to attack the Bolsheviki also helps to interpret the announcement made to the French Chamber of Deputies by M. Pinchon last Sunday that Perm had been captured by an anti-bolshevik army of Russians. It may be so; and it may even be that it was, as he declared, a tremendous victory, with 16,000 Bolsheviki killed or captured, though all news from Russia has to be scaled down. The victory, it is true, was suspiciously timely with a new adventure

in the Ukraine being floated against protest, and M. Pinchon has not shown as an authority on events in Russia. But for the present we have no evidence one way or another. Vladivostok doubles the number of prisoners, but is a remote source. The point of interest is that the French foreign secretary made a virtue of the Russians acting alone and hid from the French Parliament the news, undoubtedly barred in France by the censorship, that the Czechs had been disgusted by the setting up of a dictatorship and had refused to fight for it. When the people of the entente countries learn how they have been tricked we must expect a strong revulsion of feeling, not in favor of the Bolsheviki, but in favor of stopping stupid interference of this sort and letting Russia settle for itself what its people want.

One of the unfortunate consequences of the French Government to continue to harass the Bolsheviki with minor military operations is that it gives excuse for a continuance of the censorship which has made possible the blundering policy of the allies. In our own country, now that the war has been won, we may look for a quick return to the publicity that Americans like. The only possible reason for secrecy would be that the allies were managing things badly and antagonizing the Russian people, and if this be true the fact should not be hidden, but set right. If the Bolsheviki are too strong to be overthrown, the question will arise how they are to be treated. Are they to be declared outlaws or recognized as representing Russia or put on the waiting list? These are questions depending on facts, and the democracies of France, Great Britain, and the United States are perfectly competent to decide when a foreign government has earned recognition. The public approved of the refusal to make a commercial treaty with the Czar while the passports of American Jews were dishonored; it can be trusted to judge Russia fairly if the facts are fully and impartially presented. We may confidently expect the administration soon to take measures to that end.

It was suggested last week that to get a coherent policy the allies would have to decide whether the menace of Bolshevism was a matter of force or of ideas. Confusion upon this point has been growing since, then, for the reason that as soon as the rumor got about that the allies had decided not to invade Russia the interventionists began to "play up" the dire peril of a Russian invasion. While they still hoped for intervention Trotzky's army was derided as fit only for Falstaff; as soon as the allies decided that it was indiscreet to attack the Bolsheviki the opposite tactics were adopted; the scare-crow Red Guard became overnight a superb army, millions strong, officered by German generals, and setting out unprovoked to conquer Germany and France by sheer force. One day Trotzky was packing up and a corporal's guard of allies, one might think, would do to take Moscow and save civilization ; the next day the allies were invited to beat back countless Muscovite hordes, warlike, disciplined, well armed, well clothed, fanatically devoted to the propaganda of Bolshevism.

For this mystery it is hard to find a parallel except in Russian finance. Financial history contains nothing more entertaining than the theory that the inexplicable strength of the Bolsheviki is due to German gold and that they arc now using this ill-got wealth to bribe Germany in turn, a financial operation equaled only in the town where the people lived by taking in each other's washing. But the past week has seen a diverting addendum to this theory in the explanation that the strength of the Bolsheviki is due to their control of the printing press. With that remarkable engine, it seems, they can produce money in unlimited quantities, whereas Dictator Kolchak at Omsk is repining because many bales of paper money, neatly printed in America, are held up at Vladivostok pending developments. Apparently, then, the only hope of the anti-Bolsheviki is that the printing press at Moscow may break down from overstrain. For, while that operates, they are able to bribe all Russia and to maintain a huge army by paying in paper rubles, to be sure, fantastic wages to private soldiers, with surplus enough to bribe Germany, too.

No labored explanation of the hold of the Bolsheviki on Russia is needed, however, if we assume (1) that despite denials it has steadily been growing in strength, and (2) that the allies by a wrong policy have played into the hands of the Lenine government. Of the actual facts we as yet know little, but by inference from known conditions we may correct some of the wildest distortions of the truth. We may conclude off-hand that as an offensive force the Boisheviki are not formidable. They have small arms enough, but no such equipment of artillery and other elaborate apparatus as would nowadays be required for such a crusading war as the French Revolution waged when goaded

*into it by attacks very much like those which the French are now
making. Nor have the Russians, who got thoroughly weary of the war,
shown an aggressive spirit. Whatever the danger of Bolshevism may
be, it is not at present the danger of a march to the Rhine.*

*It is to be assumed, too, that Russia is weak because it is a tightly
blockaded country. It can produce food enough and can manufacture
certain articles, but it is industrially backward and has long been cut
off from foreign trade. It must therefore be progressively impoverished
and the less capable of attacking the right and well-equipped nations
of the west, even if the Russians should be seized with a warlike spirit,
which at present they fortunately do not show.*

*The military peril, then, may be dismissed except as it affects Russia
itself, including, of course, Siberia and the territories which the enemy
seized and are now evacuating. Of these the chief is Ukraine, and
there, as we have seen, the French are combating a separate revolution,
which is at present anti-Bolshevist, though under attack by foreign
armies supporting the dictator, Denikin, it might in self-defense join
forces with the Lenine government.*

*Next in importance to Ukraine come the Baltic Provinces, from
which the German armies are retreating, followed by a wave of
Bolshevism. This is not wholly Russian in character, however, because
the Bolshevist movement in Russia has found no stronger supporters
than the great number of revolutionists, specially Letts, who were
driven out of these Provinces by the Government, and having no home
or job in Russia, very naturally entered the red guard; to suppose them
to be mere mercenaries, as has been alleged, is part of the bribery myth
already referred to. They are, in fact, ardent revolutionists, and it is
to be presumed that they make the main force of the armies which are
recapturing their countries as the Germans retire.*

*The case of Lithuania is specially interesting, because the issue
there is complicated by Polish imperialism. Poland in its grand days
held Lithuania in a manner curiously like that in which the Austrian
Empire held the old kingdom of Bohemia, and great Russia the
Ukraine, a union for defense was perverted by successive encroachments
into subjugation. Now that the Poles have won their own freedom,
at which we are all delighted, they are seeking also to reclaim their
lost empire, and their plea for invading Lithuania is the necessity of
defending it against the Bolsheviki. It is a convenient plea, but the
Russian doctrine of self-determination has got so strong a hold in
eastern Europe that a Polish invasion of non-Polish territory is sure to*

make mischief. Moreover, we are beginning to hear that the Poles, too, are being "infected."

This infection of all the armies, not excluding the Czechs, which have fought the Bolshevik!, is a very curious phenomenon, yet not inexplicable. It can be understood if we assume (1) that the common people in all countries are averse to further slaughter for imperialistic ends, and (2) that the allied Governments by their intense secrecy as to what they are up to have left the people at the mercy of Bolshevist propaganda. For the Bolsheviki have a very definite propaganda. They assert that the Germans and the allies are two of a kind, that both have been fighting for conquests, for annexations, for indemnities. They have, moreover, certain documentary evidence in the shape of the treaties found in the Czar's archives. This sort of propaganda, taken with the predilection of the allied diplomats for supporting Czarist and reactionary Russians, is very unsettling to the average soldier, so long as his own Government keeps him in the dark. It is the issue of imperialism, therefore, which at present makes Bolshevism a menace. The menace would be greatly reduced by a just peace.

[From the *Springfield Republican*, Jan. 12, 1919.]

The feeling grows that the whole problem, in fact, of dealing with the revolution has been badly handled, and no great surprise was caused by the announcement Tuesday of the decision of the British Government against further military intervention in Russia. Assurance was given that there was no purpose of sending more troops, and that steps had been taken toward recalling part of the 20,000 men already in Russia. No statement of the reasons for this change of policy or as to the course to be followed in the future has yet been made. To some extent the decision may have resulted from the discussions held while Mr. Wilson was in London, for Russia was uppermost at that time, and it was understood that the general question of intervention was to be among the first topics to come up. But weight must also be given to the strong feeling in Great Britain against a protraction of the war; victory has been won and the people want peace. In particular the men who were conscripted to fight Germany are clamoring for demobilization. They do not want to police the continent indefinitely, and still less do they want to be ordered abroad for a war against Russia, which is what intervention means at the present time. We need not overstress reports that fear of the spread of Bolshevism in England led to the decision, but unquestionably it is now realized that intervention as the allies have actually applied it has been a

*potent means of fanning the flames of revolution—fanning them, be
it noted, toward the west.*

*What might, perhaps, have been accomplished by really adequate
intervention there has been no occasion to discuss in these reviews,
because at no time have the allies had resources and facilities for
adequate intervention. While the war lasted they were restricted
to roundabout routes of access, and when victory opened the Black
Sea and the Baltic revolution had spread so widely as to make what
might happen at Moscow or Petrograd an affair of minor moment.
At present, with the temper of Europe what it is, even a strong international campaign against the Russian soviets might involve more
risk than a statesman would care to run. Even France, where feeling
is most bitter against the new regime in Russia, there is no willingness to send an army to Moscow, though the French Government
persists in its fatuous course of fighting the revolution in Ukraine.*

*To say that intervention has failed is not to say that the Bol-
sheviki have won. Their government may collapse any day; and
though unverified, the report is not incredible that Lenine has been
arrested by Trotsky on the charge of readiness to compromise. Of the
two Trotsky appears to be the more violent and the more vindictive;
there seems to be some ground for the assertion from Russian sources
that the savage reprisals for the attempt on Lenine's life would not
have been made if he had not been laid up with his wound. Lenine is
extreme enough in his theories, but seems less energetic in action than
Trotsky, upon whom has devolved a great part of the executive side of
the revolution, including the organization of the new army. Control
of that would give him an obvious advantage over his colleague if
there has been, as rumored, a clash between them. In any case the fall
of the Lenine government is more likely if the allies leave it alone
than if by invading Russia they rally millions of ignorant people to
support against a foreign foe.*

*When a bad military policy has decisively failed it is well to go
back and examine freshly the reasons assigned for undertaking it. For
the Russian campaigns no better statement can be found than in an
article by the Russian, Raul Vinogradoff, resident in London, which
was published last June, and which not improbably had a decisive
anil unfortunate effect upon entente policy. He argued:*

*"The first requirement is to create a basis for operations by es-
tablishing points d'appul for military stores and opening up lines of
contact with the Russian people. Convenient centers are indicated by*

geography and are partly equipped even now. I mean, primarily, the Murman line and Archangel; in a lesser degree Vladivostok and the Manchurian railway; in the future possibly the Persian front. As the Germans are taking advantage of their proximity to Russia by foraging freely in the cornfields of the south, Great Britain and her allies should take advantage of the command of the sea to establish military centers in the Murman and in Archangel: nor could there be any suspicion of selfish encroachment if the Japanese acted in conjunction with the Americans and British on the Siberian front * * * it would certainly be a grievous mistake on the part of the allies if they did not use the short summer season to create a strong base in northern Russia * * *. If Great Britain and France are able to spare men and material for Salonica, surely they ought to be able to send a few thousand volunteers to secure and develop their line in contact with Russia."

Two months later the allies seized Archangel by force, and the test of these theories began. It can now clearly be seen, and was suspected by close students of the war at the time, that the theories on which intervention was based were unsound in every particular.

To get at Germany the allies had to get through Russia. To get through Russia they had either to work through the de facto government or set up another; they refused to work through the de facto government, even had that been possible, therefore they could not get at the Germans until they had set up a new government in Russia. But this meant a new Russian revolution, simply as a preliminary to a new campaign against Germany, and while one may disturb an enemy by starting a revolt in his country, it is not so easy to get help from an ally by starting a civil war as a preliminary. But this was precisely what the allies undertook to do in Russia. Whether it was a legitimate thing to do or not is beside the point; it failed, because it was based on false assumptions. Not even the censorship could long hide the fact that conditions in Russia were not in the least as they were depicted in the international press, and that instead of raising Russian armies to fight Germany, the allies were rapidly drifting into war with the Russian people. That insane adventure is well abandoned, peace just now is the greatest need.

Mr. JOHNSON of California. Mr. President, the chairman of the Foreign Relations Committee in hot resentment recently replied to the strictures of the Senator from Iowa upon the corporation termed the "War Trade Board of the United States Russian Bureau." I listened with interest to his remarks then. The organization was incorporated November 6, 1918, for the purpose of accomplishing, as the Senator from Nebraska said:

The economic penetration of Russia for the purpose of bringing relief to a country that was fairly weltering in distress and misery, many of whose social and political crimes were growing out of that misery.

Five million dollars was put aside for the purpose described by the Senator from Nebraska as—

Friendly penetration, by the sending of American products into Russia and the bringing of Russian products out of Russia in exchange for them, to promote trade and create a prosperity that could not come about from natural causes.

All efforts ceased with the signing of the armistice, because, as the Senator from Nebraska said:

"We have come to the end of the war substantially."

What a marvelous situation! We prepared in the early days of November last to supply Russia with what she needed and to take from her what she could sell—a friendly economic penetration—and to aid the Russian people. How could it be done except in conjunction with the very people who were in control of Russia? We see, therefore, an elaborate scheme initiated by our Government, without consultation at all, apparently, for economic cooperation with Russia; and yet, when the offers in writing for economic cooperation were made again, and again, during 1918, by the very people with whom we would have to deal in November, our Government, apparently, would not respond or reply or have aught to do with the offers, because of the wickedness of their authors; and then, after repulsing these offers with contemptuous silence, prepares itself with $5,000,000 to do the very same thing. It ceased its efforts with the armistice. Just think of it! On the 6th day of November, 1918, in the language of the Senator from Nebraska, we had started on the enterprise with Russia of "bringing them shoes and food and clothes and the things they were without." Five days later we ceased our efforts and brought the Russians instead starvation and bullets and bayonets. Our efforts at charity and benevolence ended, because, as the Senator said:

We have come to the end of the war substantially,

and I presume, for the same reason, because we had come to the end of the war substantially, and would not deal in friendliness

and charity, we began to invade and to shoot and to starve and to
kill. We "had come to the end of the war substantially," and we
could not, therefore, economically penetrate Russia, but we could
shoot Russians after we "had come to the end of the war substantially."

Mr. President, no one has less sympathy than I with the Bolsheviki of Russia, who there ruthlessly control with a hideous class
tyranny. None will more scornfully reject their grotesque doctrine.
But I will not permit my feelings for the men or their formulae to
blind me to our own wrongdoing, nor will I cloak our wrong with
hypocritical denunciation. During the war it became fashionable
to call all who disagreed with any governmental policy pro-Ger-
man. Now the fashion has changed; and any man who will not
accept the wrongful edict of intrenched power is by that token a
Bolsheviki. In making the world safe for democracy we have put
our intellects in chains; and one of our first tasks with ourselves is
to unlock the prisons in which we have confined our brains. I read
of great statesmen of ours saying "Shoot them down" and others
"Hang them." You can not shoot or hang a state of mind. When it
becomes by open expression treasonable, under the law of the land
you may punish it. But if men in high places imagine by invoking
lawlessness against lawlessness they can make the world better,
all history denies them. I think too well of my country to believe
for one instant the doctrine of Bolsheviki Russia can ever find a
foothold here. My faith in the Republic and in our people, in our
democracy, will not permit me to be frightened by the fantasy of
madmen elsewhere. I think I can with equanimity observe the ser-
vile part of the press apply to me for this speech the now familiar
epithet of Bolsheviki. Its indiscriminate application is illustrated by
a New York administration paper designating prohibitionists as the
Bolsheviki of America.

My appeal to-day will find no response with those newspapers
and great men who preach anarchy when they demand killing
and hanging out of hand; but it will have its answering approval
with the inarticulate mass who ask but justice and the same honesty in governments as in men, it will find its echo in the hearts of
the common folks whose sons and husbands in frozen Russia are
paying the price of our Government's wrong and broken faith, and
I am content.

Why did we enter Russia? I answer, for no very good reason;

and we have remained for no reason at all.

What is our policy toward Russia? I answer we have no policy. We have engaged in a miserable misadventure, stultifying our professions, and setting at naught our promises. We have punished no guilty; we have but brought misery and starvation and death to the innocent. We have garnered none of the fruits of the victory of war, but suffer the odium and infamy of undeclared warfare. We have sacrificed our own blood to no purpose, and into American homes have brought sorrow and anguish and suffering.

<u>Bring the American boys home from Russia.</u>

Explicit iste liber, scriptor sit crimine
liber, Christus scriptorem custodiat ac
det honorem

Ὥσπερ ξένοι χαίρουσιν ἰδεῖν
πατρίδα, οὕτως καὶ οἱ γράφοντες
ἰδεῖν βιβλίου τέλος

श्रीकृष्णार्पणमस्तु

書成矣，感盡天地

סלוע ארוב לאל חבש סלשנו סת

"of making many books there
is no end; and much study is a
weariness of the flesh"
- Ecclesiastes 12:12

BULKINGTON BOOKS